Rita J. Terry

FOOD PREPARATION STUDY COURSE

Quantity Preparation and Scientific Principles

THIRD EDITION

Food Preparation Study Course

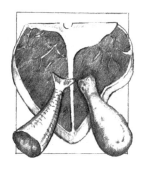

QUANTITY PREPARATION and SCIENTIFIC PRINCIPLES

THIRD EDITION

IDA IOWA DIETETIC ASSOCIATION

Lynne E. Baltzer, Ph.D., R.D., L.D.

Shirley A. Gilmore, Ph.D., R.D., L.D.

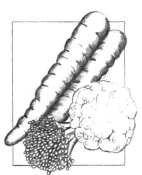

IS IOWA STATE UNIVERSITY PRESS / AMES

Research Associates: Don M. Paulson, M.S., R.D., L.D.; In-sook Lee, M.S.

Reviewers: Anne Kriener Blocker, R.D., L.D.; Julie Metcalf Cull, R.D., L.D.; Marlys Lane, R.D., L.D.; and Monica Lursen, R.D., L.D.

Content Reviewer: Jeanette Bohnenkamp, M.S.

Language Consultant: Rosalind Engle, Ph.D.

Lynne E. Baltzer, Ph.D., R.D., L.D., is an associate professor of Hotel, Restaurant, and Institution Management at Iowa State University, where she teaches quantity recipe development and courses in foodservice management, computer technology, catering, and international cuisine.

Shirley A. Gilmore, Ph.D., R.D., L.D., is an associate professor of Hotel, Restaurant, and Institution Management at Iowa State University, where she is the principal instructor of Personnel Management and also has taught Quantity Food Production Management Experience.

Graphic artist: Tom Hiett, graphic specialist, Media Resources, Iowa State University

©1971, 1985, 1992 Iowa State University Press, Ames, Iowa 50010
All rights reserved

Authorization to photocopy items for internal or personal use, or the internal or personal use of specific clients, is granted by Iowa State University Press, provided that the base fee of $.10 per copy is paid directly to the Copyright Clearance Center, 27 Congress Street, Salem, MA 01970. For those organizations that have been granted a photocopy license by CCC, a separate system of payments has been arranged. The fee code for users of the Transactional Reporting Service is 0-8138-0806-5/92 $.10.

∞ Printed on acid-free paper in the United States of America
Third edition, 1992
Second printing, 1993

Library of Congress Cataloging-in-Publication Data

Baltzer, Lynne E.
    Food preparation study course : quantity preparation and scientific principles / Lynne E. Baltzer, Shirley A. Gilmore.—3rd ed.
       p.   cm.
    "Iowa Dietetic Association."
    Rev. ed. of: Food preparation / prepared by Louise Dennler. 2nd ed. 1985.
    Includes bibliographical references and index.
    ISBN 0-8138-0806-5 (alk. paper)
    1. Quantity cookery. I. Gilmore, Shirley. II. Dennler, Louise.
Food preparation. III. Iowa Dietetic Association. IV. Title
TX820.D45 1992
641.5'7—dc20
                                                                                                                                                              92-5530

CONTENTS

Foreword, ix
Preface, xi
How to Use the Study Course, xii

1. **RECIPE STANDARDIZATION, 3**
   Developing Standardized Recipes, 3
      Selecting a Recipe Format, 4
      Writing a Recipe, 6
      Recipe Names and Filing Systems, 7
      Recipe Header and Body, 7
      Weights and Measures, 8
   Recipe Standardization Procedures, 11
      Using the Recipe Standardization Procedure, 12
      Selecting a Yield for Testing, 13
      Planning for Evaluation, 13
      Planning for Production, 13
      Preparing the Recipe, 13
      Portion Control, 14
   Completing the Standardization of a Recipe, 18
      Evaluating the Recipe, 19
      Identifying Solutions to Quality Problems, 19
      Final Testing of the Recipe, 19
   Adjusting Recipes, 19
      Adjustment Methods, 19
   Summary, 22
   Learning Activities, 22
   Review Questions, 24

2. **FRUITS, 25**
   Understanding Fruits, 25
      Classification, 25
      Structure and Composition, 26
   Selection, 26
      Fresh Fruits, 26
      Processed Fruits, 28
   Storage, 29
      Fresh Fruits, 29
      Processed Fruits, 29
   Preparation, 30
      Fresh Fruits, 30
      Processed Fruits, 31
      Cooking Methods, 32
   Summary, 33
   Learning Activities, 33
   Review Questions, 35

3. **VEGETABLES, 38**
   Understanding Vegetables, 38
      Characteristics, 38
      Classification, 39
   Selection, 40
      Fresh Vegetables, 40
      Processed Vegetables, 41
   Storage, 42
   Preparation, 43
      Fresh Vegetables, 43
      Processed Vegetables, 45
   Controlling Quality Changes During Cooking, 47
      Effects of Cooking on Texture, 47
      Effects of Cooking on Color, 48
      Effects of Cooking on Flavor, 49
      Effects of Cooking on Nutritional Content, 50
   General Rules of Vegetable Cooking, 50
   Potatoes, 50
      Preparation, 51
      Storage, 51
      Cooking, 51
   Summary, 52
   Learning Activities, 52
   Review Questions, 56

4. **SALADS, 59**
   Understanding Salads, 59
      Role of Salad in the Menu, 59
      Standards for Salad, 59
      Parts of a Salad, 60
   Types of Salads, 60
      Fruit Salad, 61
      Vegetable Salad, 61
      Molded Gelatin Salad, 61
      Protein Salad, 61
   Preparation, 62
      Preparing Greens, 62
      Preparing Molded Gelatin Salad, 62
      Preparing Protein Salad, 63
      Tools Used in Preparing Salad, 63
   Plating Salads, 64

Garnishes, 65
Salad Dressings, 65
   Types of Salad Dressings, 65
   Selecting Salad Dressings, 66
Summary, 66
Learning Activities, 67
Review Questions, 73

5. STARCHES, SAUCES, SOUPS, CEREALS, AND PASTAS, 75

Starch-thickened Foods, 75
   Puddings and Pie Fillings, 78
   Sauces, 79
   Gravies, 80
   Cream Soups, 80
Cereals, 80
   Cooking Methods for Cereals, 81
Rice, 82
   Cooking Methods for Rice, 82
Pasta, 83
   Noodles, 84
   Cooking Methods for Pasta, 84
   Holding Pasta, 85
Summary, 85
Learning Activities, 86
Review Questions, 90

6. MILK AND CHEESE, 92

Milk, 92
   Nutritional Values of Milk, 92
   Milk Processing, 93
   Cooking with Milk, 94
Cheese, 98
   Types of Cheese Products, 99
   Cooking with Cheese, 102
Summary, 102
Learning Activities, 102
Review Questions, 105

7. MEATS, POULTRY, FISH, AND ENTREES, 106

Meats, Poultry, and Fish, 106
   Nutrient Content, 106
   Composition, 106
   Meat Products, 108
   Poultry Products, 109
   Fish and Seafood Products, 109
   Storage, 110
   Meat Preparation and Cooking, 111
   Poultry Preparation and Cooking, 113
   Fish Preparation and Cooking, 115
Leftovers, 115

Sandwiches, 116
   Sandwich Production, 116
Summary, 117
Learning Activities, 117
Review Questions, 123

8. EGGS AND EGG PRODUCTS, 125

Uses of Eggs, 125
   Thickening Agent, 125
   Binding Agent, 125
   Emulsifying Agent, 126
   Leavening Agent, 126
   Coloring and Flavoring Agent, 126
Egg Quality and Size, 126
Purchasing and Storing Eggs, 127
   Purchasing, 127
   Storage, 127
   Market Forms of Eggs, 127
Cooking Eggs, 130
   Methods of Cooking Eggs, 131
   Foams, 133
Summary, 134
Learning Activities, 135
Review Questions, 141

9. DOUGHS, BATTERS, AND PASTRIES, 143

Bread Doughs and Batters, 143
   Yeast Breads, 143
   Quick Breads, 148
Pancake and Waffle Batters, 150
Cakes and Cookies, 150
   Cakes, 150
   Cookies, 153
Pastries, 153
   Pie Dough, 153
   Puff Pastry, 154
Summary, 154
Learning Activities, 155
Review Questions, 160

10. BEVERAGES AND CONVENIENCE FOODS, 162

Beverages, 162
   Coffee, 162
   Tea, 164
   Milk Beverages, 164
   Carbonated Beverages, 165
Convenience Foods, 165
   Types of Convenience Foods, 165
   Storing and Preparing Frozen Convenience Foods, 165
   When to Use Convenience Foods, 166

Summary, 167
Learning Activities, 167
Review Questions, 170

11. MICROWAVE COOKING, 172
How Microwave Ovens Work, 172
    Oven Wattage, 174
    Microwave Cooking Principles, 175
    Cooking Materials and Utensils, 176
    Foods Not Recommended for Microwaving, 178
Microwave Uses in Quantity Food Production, 178
Microwave Safety, 179
Summary, 179
Learning Activities, 185
Review Questions, 186

12. MENU PLANNING, 187
Factors Affecting Menu Planning, 187
    Clientele, 187
    Availability of Food, 191
    Availability of Equipment and Arrangement of Physical Facilities, 191
    Personnel Skills, 191
    Money Budgeted for Food, 191
    Style of Service, 192
The Cycle Menu, 192
    Advantages of the Cycle Menu, 192
    Disadvantage of the Cycle Menu, 192
Procedures for Good Menus, 193
    Planning Menus, 193
    Writing Menus, 194
    Evaluating Menus, 195

Summary, 196
Learning Activities, 196
Review Questions, 197

13. FOOD PURCHASING, RECEIVING, AND STORING, 199
Purchasing, 199
    Government Regulations about Food Quality, 199
    Brands, 200
    The Purchaser's Responsibilities, 201
    Centralized Buying, 201
    Cooperative Buying, 201
    Information Needed by Purchasers, 201
    Choosing Purveyors, 202
    Methods of Purchasing, 202
    Food Quality, 203
    Determining Quantity, 204
    Writing Specifications, 204
    Purchase Orders, 205
Receiving Food, 206
Storing Food, 206
    The Dry-Storage Area, 207
    Refrigerated or Freezer Storage, 207
    Inventory Control, 208
Summary, 208
Learning Activities, 209
Review Questions, 212

Food Preparation Terms, 214
Answers to Chapter Review Questions, 216
References, 218
Index to Recipes, 219

FOREWORD

"It is good food and not fine words that keep me alive."

—Molière

    The enjoyment of good food is a common link among all people regardless of place or circumstance. *Food Preparation Study Course, Third Edition,* is designed to ensure that more good food will be available for more people to enjoy!

    This guide is intended for mentored self-study use by persons employed in the foodservice industry, and as a training tool for dietitians, dietary consultants, and managers. It contains important lessons relating to menu planning; food purchasing, receiving, and storage; recipe standardization; and preparation of all types of food.

    The original study course was developed in 1971 by Louise Dennler, R.D. At that time she was a member of the nutrition staff of the Iowa State Department of Health and was very much aware of the need of foodservice employees in health care facilities for more information on proper procedures for preparing food.

    Subsequent editions of this course were revised by members of the Iowa Dietetic Association, and the course has since become a project of that association. Proceeds go into continuing provision of educational materials for the foodservice employees in Iowa health care facilities and elsewhere.

Judith Klopfenstein, M.S., R.D., L.D.
Iowa Dietetic Association President, 1991–1992

PREFACE

Food preparation in foodservice organizations involves changing foods received from the supplier into finished products for service. The heart of quantity foodservice is food preparation, which is an art as well as a science. Knowledge, skill, and creativity are required to make food attractive, flavorful, and interesting to the client.

It is important for foodservice employees to understand the goals of the foodservice department when preparing food to be served to the clients. It also is important for them to understand the functions of menu planning, purchasing, receiving, storage, recipe standardization, and preparation of all types of foods.

Food Preparation Study Course, Third Edition, provides comprehensive information for foodservice employees. The study course has thirteen chapters, each with learning activities that will help the student apply the food preparation theories and methods discussed. The principles also will help explain why certain things happen as they do in food preparation, the many chemical and physical changes that affect color, flavor, texture, and nutrition of foods.

The overall objectives of the *Food Preparation Study Course* are for students to be better able to

- Plan, select, and prepare foods for the clients, guests, and personnel in a foodservice organization

- Understand how chemical and physical changes affect food preparation

- Use standardized recipes and methods to prepare foods in ways that

  - Obtain uniformly good products
  - Maximize nutritional value
  - Improve or retain flavor and color
  - Provide texture variety
  - Maintain temperature control
  - Control portion size
  - Create a pleasing appearance

Terms specific to food preparation are defined in the section Food Preparation Terms at the back of the book. Knowing the meanings of common terms makes it possible to understand standardized recipes and follow the directions.

How to Use the Study Course

The *Food Preparation Study Course* is designed to provide information and practical experience in food science and preparation for quantity food. This study course should be used in combination with direction from a professional who is qualified to guide the learning activities. Each chapter includes text, learning activities, and review questions. Additional references are suggested for further reading.

Students should read the text for a chapter and then complete the learning activities, either during the class time or in their own facilities. It may be necessary to refer to the text when completing the learning activities. Students should be encouraged to apply what they learn to their jobs.

After completing the learning activities, students should carefully read and complete the review questions for that chapter. Their answers can be compared with those provided, and any incorrect answers should be reviewed by reading the text again.

NOTE TO INSTRUCTOR

Food Preparation Study Course, Third Edition, is written to be completed in a twelve-hour course as part of the ninety-hour supervisor's education program in Iowa. The instructor should adapt the study course to meet students' needs. The chapters may be covered in any order the instructor prefers. Some instructors will choose not to cover some of the chapters, but rather encourage the students to use the material as resource information.

An alternative order for the chapters may be

1. Menu Planning
2. Recipe Standardization
3. Food Purchasing, Receiving, and Storing
4. Meat, Poultry, Fish, and Entrees
5. Eggs and Egg Products
6. Milk and Cheese
7. Starches, Sauces, Soups, Cereals, and Pastas
8. Vegetables
9. Fruits
10. Salads
11. Doughs, Batters, and Pastries
12. Beverages and Convenience Foods
13. Microwave Cooking

FOOD PREPARATION STUDY COURSE

Quantity Preparation and Scientific Principles

THIRD EDITION

1. RECIPE STANDARDIZATION

Foodservice operations are evaluated on the cost-effective production of high-quality food in the correct quantity to meet the client's needs. The necessary equipment and ingredients and trained employees must be available to meet this standard. The work must be planned to make sure that menus include appropriate foods and that purchasing methods ensure the needed foods are on hand. Production methods must yield food of the correct quantity of desired quality every day. The recipe is particularly important in meeting these high standards. Regardless of other resources available, the quality of the recipe is a deciding factor in the success of the foodservice.

Quality recipes for quantity foodservices must provide the information needed to produce a desirable product in the correct quantity, every time it is made. The recipe standardization process makes sure a recipe suits the needs of the foodservice. The process makes the recipe suitable for

- Production equipment available
- Quality desired
- Type and quality of food purchased
- Usual quantity produced
- Type of food production and service systems

Standardized recipes are very helpful in training new cooks or when a substitute or relief worker must do the preparation. The changes in personnel that happen in all foodservices will not have as much effect on production when standardized recipes are relied upon. Purchasing is also more efficient. There is no guesswork about amounts to purchase, and it is easier to calculate costs of menu items. Using a standardized recipe, the exact quantity of ingredients can be ordered and the number of portions that will be prepared can be accurately predicted.

DEVELOPING STANDARDIZED RECIPES

Recipes are available from many sources. They can be obtained from

- Employees
- Clients and their families
- Foodservice magazines
- Cookbooks used at home
- Cookbooks with large-quantity recipes
- Food manufacturers and vendors
- Food commodity groups like the beef, pork, and egg councils
- Government agencies
- Other nursing facilities, hospitals, and schools

Some of these sources provide tested quantity recipes, but many are untested. A *tested* recipe means it has been made in large quantity and evaluated for flavor, standard production methods, and use with foodservice equipment. A tested recipe also can be counted on to produce the same quantity and quality any time it is prepared. Quantity recipes published by foodservice magazines, cookbooks, manufacturers, vendors, commodity groups, and government agencies usually are tested.

Recipe standardization is the process of making a recipe suit a specific purpose in a particular foodservice operation. Tested recipes are the easiest to standardize because they are already written using quantity food production methods and equipment. However, a tested recipe is *not* standardized until it is used in a specific facility. It is not possible to purchase standardized recipes. Tested recipes can be obtained, but the standardization must occur in each production unit. When a recipe is standardized for one facility, it would have to be standardized again for use when there are any changes in equipment or production methods, as well as major changes in food products purchased.

Selecting a Recipe Format

The first step in standardizing recipes is to decide a format that will be best for the facility. Using the same format for all recipes in the organization makes them easier to use. Everyone who uses the recipes—purchaser, supervisor, menu planner, cook, and dietitian—will know where to look for needed information on each recipe. A format should be selected that is

- Easy to read from at least 18 inches away
- Large enough so most recipes will fit on one page

Several recipe formats have been used by quantity foodservices. Each has advantages and disadvantages. The most common types are the standard form, the action form, and the block form. The block form is the best format to use for quantity food production.

THE STANDARD FORM. The standard form lists all the ingredients at the top of the form in the order they will be used. The steps in the preparation procedure are listed below the ingredients in the order they will be completed, as in the Oven-fried Chicken recipe. This form allows easy listing of the quantity of ingredients for different yields. But because the cook must constantly look in two places, the ingredient list and the method steps, mistakes are more likely than when the block form is used. It is possible to miss a step or put an ingredient in twice when looking back and forth between the two lists.

RECIPE STANDARDIZATION

STANDARD FORM RECIPE FORMAT

Oven-fried Chicken

Portion Size: 1 chicken quarter

Baking temperature: 375 F
Baking time: 60 min
Equipment: Deck oven

100 portions	125 portions	Ingredients
100 quarters	125 quarters	Chicken fryers, 2.5–3 lb, quartered
2 lb	2 lb 8 oz	Flour, all-purpose
¼ c 3 TBSP	½ c 2 tsp	Salt
2 TBSP 2 tsp	3 TBSP 1 tsp	Pepper, white
2 TBSP 2 tsp	3 TBSP 1 tsp	Paprika
1 lb	1 lb 4 oz	Margarine, melted

1. Wash and trim chicken.
2. Drain thoroughly.
3. Combine flour and seasonings.
4. Coat chicken lightly with flour mixture, shaking off excess.
5. Place chicken, skin side up, on lightly greased sheet pans.
6. Brush top of each chicken quarter with melted margarine.
7. Bake at 375 F for 1 hr or until chicken is browned and tender.

ACTION FORM. The action form recipe format also lists ingredients and procedures in the order they are used. The information is presented in paragraphs that include both ingredients and procedures. This form is difficult to use, as can be seen in the Oven-fried Chicken example.

ACTION FORM RECIPE FORMAT

Oven-fried Chicken

Wash, trim, and drain 100 chicken quarters. While the chicken is draining combine 2 lb all-purpose flour, ¼ c + 3 TBSP salt, 2 TBSP + 2 tsp white pepper, and 2 TBSP + 2 tsp paprika. . .

THE BLOCK FORM. The best format to use for quantity food production recipes is the block form. As the example shows, the ingredients are listed in the order used on the left side and the procedure for each ingredient on the right. Ingredients are grouped with procedures and divided by solid lines to make the recipe easier to read. This form also makes it easier for the cook to identify what to do with each ingredient and the correct preparation sequence. The column on the left is for amounts in recipes modified to a larger or smaller number of portions.

BLOCK FORM RECIPE FORMAT				
Oven-fried Chicken				
.... Yield: 100 portions .... Portion size: 1 quarter			Baking temperature: 375 F Baking time: 60 min	
Amount	**Ingredients**	**Amount**	**Procedure**	
....	Chicken, fryers, 2½–3 lb, quartered	100	1. Wash and trim chicken. 2. Drain thoroughly.	
....	Flour, all-purpose............... Salt..................................... Pepper, white...................... Paprika...............................	2 lb ¼ c 3 TBSP 2 TBSP 2 tsp 2 TBSP 2 tsp	3. Combine flour and seasonings. 4. Coat chicken lightly, shaking off excess. 5. Place chicken, skin side up, on lightly greased sheet pan.	
....	Margarine............................	1 lb	6. Brush top of each chicken quarter with melted margarine. 7. Bake at 375 F for 1 hr or until chicken is browned and tender.	

Source: *Standardized Quantity Recipe File* (1971).

Writing a Recipe

A recipe should contain all information necessary for using it in menu planning, purchasing, scheduling labor and equipment, and preparing and serving the food. A well-written recipe usually includes the following:

- The name of the recipe
- The recipe group, such as Beef Entree
- The recipe file code number
- The total quantity produced
- The number of portions produced
- The portion size and/or weight
- The ingredients and quantity in weight, measure, volume, or count
- The procedures and approximate times for preparation steps
- The cooking times and temperatures
- The mixing and cooking equipment
- The portioning instructions and utensils
- The serving temperature

A recipe may also contain portion cost, quantities to purchase, a list of meals on the menu cycle when it is served, a history of portions prepared, and a description of the finished product.

When writing recipes, 8½ × 11″ paper works best. This size is large enough for most recipes to fit on one page and is more easily typed or word-processed than smaller cards. Page-size plastic sheet protectors can be purchased to protect the recipes from spills.

Print size is very important for recipes. It should be large enough to read when the recipe is posted on the cook's counter because it is more efficient not to pick up the recipe to read each step. The largest size available on a word processor or computer that will allow the recipe to be printed on a single page should be used.

Recipe Names and Filing Systems

RECIPE NAMES. A system for naming recipes and organizing the recipe file helps the dietitian, manager, and production workers find particular recipes. In nursing facilities recipe names should describe the main ingredient and/or be a name familiar to the clients. It should be as appetizing as possible without being too wordy. Sometimes the names listed on the menu for clients may be different from the recipe name. For example, a recipe for beef stew might be called Iowa Beef Stew with Harvest Vegetables on the menu.

RECIPE FILING SYSTEMS. Many systems are used for categorizing and filing recipes. The two systems that are most successful are the menu cycle and the food group category. Which system is used in a specific foodservice organization depends on the preference of the manager and dietitian. The most important step is consistently using the same filing system.

Menu Cycle System. The menu cycle system files recipes together for each meal in the menu cycle. Each recipe is labeled at the top with the cycle week, day, and meal. The file contains a copy of the recipe in each meal where it will be served. Each copy can be already adjusted to the portions usually required for the meal. This makes it handy to get the recipes needed for one day and return them later. However, when a recipe is changed or a new recipe is substituted for an old one, the change must be made on each copy every time the recipe is used in the menu cycle. Sometimes each category is color-coded with a stripe along the top or by using various colors of paper to make sorting recipes filed by either of the two methods even easier.

Food Group Category System. When a food group category system is used, only one copy of each recipe is needed. In this system, the food group is listed at the top of the recipe. Each recipe is numbered within each group to make sure they are always filed in the same order. The food groups used for the file should be similar to the menu categories for the facility. A larger number of groups makes it easier to locate a specific recipe in the file, but a smaller number of groups speeds putting the recipes back in the file. The following lists shows some of the groups that might be used in the system.

- Entrees
    - Beef
    - Pork and Veal
    - Poultry
    - Fish and Shellfish
    - Casseroles
    - Egg and Cheese
- Breads
- Vegetables
- Desserts
    - Cakes, Cookies, and Bars
    - Pies, Cobblers, and Crisps
- Soups
- Sauces and Gravies
- Pasta and Rice
- Potatoes and Dressings
- Sandwiches
- Salads
    - Vegetable
    - Fruit
    - Gelatin
- Beverages
- Miscellaneous

Recipe Header and Body

RECIPE HEADER. The top portion of the recipe is called the *recipe header*. It contains important information needed in planning for production. Each recipe header should have the following information:

- Recipe name and menu name
- Yield in portions and total weight and/or pans
- Cooking equipment, time, and temperature
- Portion size and portioning utensil
- Recipe group name and number and recipe number

Most recipes are arranged like the block form Oven-fried Chicken recipe. The recipe name is at the center top, yield and portion information on the left, and cooking information on the right. The recipe group and number are at the upper right or left. A solid line dividing the header from the body of the recipe helps to locate needed information.

MAIN BODY. The body of the recipe contains the ingredients, quantity, and preparation procedures.

The ingredients are listed in the order used and grouped with the directions for using them. The ingredient entry should list the name of the food as purchased and then add the preparation needed to make it ready for the recipe. Here are three ways to list carrots in a facility's recipe, varying with the form of the purchased food.

In the first example, the cook receives the carrots whole and unpeeled:

Ingredient	Quantity	Procedure
Carrots	2 lb	1. Wash and peel carrots. 2. Cut into 1″ pieces. 3. Place in 12×20×2½″ pan. 4. Steam 6 min.

In the second example, the cook receives peeled carrots:

Ingredient	Quantity	Procedure
Carrots, peeled	1 lb 11 oz	1. Cut into ½″ slices. 2. Place in 12×20×2½″ pan. 3. Steam 6 min.

In the third example, the cook receives peeled and cut carrots:

Ingredient	Quantity	Procedure
Carrots, peeled, ½″ slices	1 lb 11 oz	1. Place in 12×20×2½″ pan. 2. Steam 6 min.

Weights and Measures

Ingredient quantities should be measured by consistent methods. Weighing ingredients is better than measuring them because it is more accurate and saves time. Workers trained to use scales will find it is faster to weigh ingredients right into the mixing bowl than first placing them in a measuring cup and then scraping them into the bowl. By measuring into the bowl, loss of food can be reduced, as well, because none is left in the measuring utensil. This is especially true for foods like shortening, corn syrup, and tomato paste.

The weight and volume relationship of all ingredients are not the same. For water 1 cup equals 8 ounces weight and 8 fluid ounces. For cake flour 1 cup weighs only 3.5 ounces. If measures must be converted to weights, the rule 1 cup equals 8 ounces can be used for liquid ingredients, such as

RECIPE STANDARDIZATION

water, liquid milk, fruit juices, and oil. The volume measurement of other ingredients should be weighed to find out the relationship. Tables 1.1 and 1.2 list some weight, volume, and measure equivalents used in food production.

Table 1.1. Common weight and volume equivalents

3 tsp = 1 TBSP = 0.5 fl oz			16 TBSP = 1 c = 8 fl oz		
2 TBSP = ⅛ c = 1 fl oz			2 c = 1 pt = 16 fl oz		
4 TBSP = ¼ c = 2 fl oz			4 c = 1 qt = 32 fl oz		
5⅓ TBSP = ⅓ c = 2.33 fl oz			2 qt = 0.5 gal = 64 fl oz		
8 TBSP = ½ c = 4 fl oz			4 qt = 1 gal = 128 fl oz		
12 TBSP = ¾ c = 6 fl oz					

Table 1.2. Equivalent weights and measures of common foods

Food	Count or Uncooked Volume	Weight or Cooked Volume
Butter and margarine	2 c	1 lb
Cheese, grated	4 c	1 lb
Eggs, medium, shelled	9 eggs	1 lb
Eggs, large, shelled	8–10 eggs	1 lb
Egg whites, medium	16 eggs	1 lb
Egg whites, large	14–18 eggs	1 lb
Egg yolks, medium	32 eggs	1 lb
Egg yolks, large	22–26 eggs	1 lb
Flour, all-purpose	4 c	1 lb
Flour, bread	4 c	1 lb
Flour, cake	4½ c	1 lb
Flour, whole-grain, coarse	3½ c	1 lb
Macaroni	4 c	1 lb
Macaroni	1 c	2¼ c
Mayonnaise	2 c	1 lb
Noodles	1½–2 qt	1 lb
Noodles	1 c	1–1¼ c
Rice	2 c	1 lb
Rice	1 c	3 c
Rice, cooked	2 qt	1 lb
Spaghetti	1 qt	1 lb
Spaghetti	1 c	1¾ c
Sugar, brown, lightly packed	3 c	1 lb
Sugar, brown, firmly packed	2 c	1 lb
Sugar, granulated	2 c	1 lb
Sugar, powdered, unsifted	3 c	1 lb
Tomato paste	1¾ c	1 lb
Walnuts, chopped	1 qt	1 lb
Yeast, dry	1 envelope or 2 TBSP	¼ oz
Yeast, compressed	1 cake	½ oz

WEIGHT UNITS. To record weights, most facilities in the United States use pounds and ounces or pounds and decimal parts of a pound. Usually the unit of weight of the scales available is the best choice for the recipe. If both types of scales are used in a food-production unit, a conversion table such as Table 1.3 can be attached to each scale.

Table 1.3. Decimal conversions for weights

oz	Part of a Pound		oz	Part of a Pound	
	Fraction	Decimal		Fraction	Decimal
0.5		.031	8.5		.531
1.0		.063	9.0		.563
1.5		.094	9.5		.594
2.0	⅛	.125	10.0	⅝	.625
2.5		.156	10.5		.656
3.0		.188	11.0		.688
3.5		.219	11.5		.719
4.5		.250	12.0	¾	.750
4.5		.281	12.5		.781
5.0		.313	13.0		.813
5.5		.344	13.5		.844
6.0	⅜	.375	14.0	⅞	.875
6.5		.406	14.5		.906
7.0		.438	15.0		.938
7.5		.469	15.5		.969
8.0	½	.500	16.0	1.0	1.000

Calculating conversions and recipe adjustments is easier using decimals rather than fractions. For example, it is easier to add 1.5 ounces and 3.75 ounces than 1½ ounces and 3¾ ounces, especially on a calculator. Decimals also take less space on the recipe form.

ROUNDING WEIGHTS AND MEASURES. Because measuring utensils and scales may not be available to measure very small amounts such as ⅛ teaspoon and ⅛ ounce, rounding is sometimes necessary. Quantity-food-production scales are usually not accurate to more than ¼ ounce. When recipe quantity is increased or decreased, the ingredient quantity may become time-consuming or impossible to measure or weigh. For example, if a recipe uses 1 pound 8 ounces of flour for 100 portions, it would require 1 pound 12.8 ounces of flour for 120 portions. Since amounts like this are difficult or impossible to weigh and/or measure with usual quantity-food-production equipment, tables have been developed to give guidelines for weights. The smaller the total quantity of the ingredient, the less rounding is appropriate. In larger quantities, rounding can be used without affecting the product quality or quantity. Table 1.4 lists guidelines for rounding ingredients. To round uneven measures and weights, find the interval on the chart containing the unrounded weight or measure of the ingredient. Then use the rule in the far right column to determine how to round each ingredient. Using the table to round the 1 pound 12.8 ounces of weight from the previous example, the cook would find that the amount should be 1 pound 13 ounces.

As can be observed from Table 1.4, the amount of rounding increases as the total quantity increases. These rounding amounts have been identified to allow recipes to be written with any quantity that it is possible to measure or weigh, yet to control the amount of rounding to prevent detrimental effects on the products.

Table 1.4. Guide for rounding weights and measures

Total Amount of Ingredient		Round to Closest
More than	Less than	
Weights		
0 oz	1 oz	¼ tsp
1 oz	10 oz	0.25 oz
10 oz	2 lb 8 oz	0.5 oz
2 lb 8 oz	5 lb	full ounce
5 lb		0.25 lb or 4 oz
Measures		
0	1 TBSP	⅛ tsp
1 TBSP	3 TBSP	¼ tsp
3 TBSP	½ c	½ tsp
½ c	¾ c	full tsp or 0.25 oz
¾ c	2 c	full TBSP or 0.5 oz
2 c	2 qt	¼ c or 0.75 oz
2 qt	4 qt or 1 gal	½ c or 1 oz
1 gal	2 gal	full cup or 0.25 lb
2 gal	10 gal	full quart or 1 lb
10 gal	20 gal	½ gal or 2 lb
20 gal		full gallon or 4 lb

RECIPE STANDARDIZATION PROCEDURES

The procedures or directions in a recipe should be as short as possible and still be clear to the reader. Using the same phrases and terms in all recipes helps workers understand what the directions mean. A set of abbreviations should be identified for use in recipes. Consistent use of abbreviations and terms in recipes like those in the following list is desirable. The explanations help decide when to use the abbreviations.

Abbreviations and Terms	Meaning and How to Use
#10 can	Use # for number or pound, not both.
10–12 lb	List lb only once.
12×20×2½"	Use " only at end of measurement.
1st speed/low	Use the same term for all recipes; low, medium, or high should be used if mixers vary in the facility.
dipper	Use "dipper," not scoop.
portions	Use "portions," not "servings."
350 F	Degree sign can be omitted.
amt	Amount
AP	As purchased
approx	Approximate
c	Cup or cups
doz	Dozen or dozens
EP	Edible portion
flat beater	Use "flat beater" or "flat paddle," but do not use both names for the same item.
flour, all-purpose	Always specify the type of flour.
gal	Gallon or gallons
hr	Hour or hours

Abbreviations and Terms	Meaning and How to Use
"	Inch or inches
lb	Pound or pounds
lg	Large
med	Medium
min	Minute or minutes
no. 10 can	Use a period after "no" to indicate "number."
oz	Ounce or ounces
fl oz	fluid ounce or fluid ounces
pt	Pint or pints
qt	Quart or quarts
sec	Second or seconds
sm	Small
sq	Square
TBSP, T, or Tbsp	Tablespoon
tsp or t	Teaspoon
wt	Weight

Using the Recipe Standardization Procedure

SELECTING A RECIPE. New recipes must always be standardized. Recipes already in use also may need to be standardized. Whether with new or existing recipes, the process of standardization is very similar.

WRITING THE RECIPE. The first step is to select a recipe to standardize. A tested recipe is usually not rewritten before it is made the first time. If a home-size recipe is selected, or if no complete written recipe is available, the recipe must first be written in correct form. Sometimes foods prepared in facilities do not have complete recipes. There may be only a list of ingredients, if that. In this case, someone must write down how the product is made. It is best to have one person observe and take notes, while a second person makes the product. The recipe is written in the format selected for the facility. The block form is usual, but the standard form might be used.

Recipe Header. The recipe header should be completed, filling in all the information that is available. Some will need to be added after the product is made, but before the recipe is made, this information can be filled in.

- Recipe name
- Yield
- Portion size, and utensil when appropriate
- Equipment

RECIPE BODY. The body of the recipe is written next.

- Convert measures to weights and record them, using the standard abbreviations and the equivalents table.
- Review the ingredient names and make them like those in the other standardized recipes in the facility.
- Compare recipe methods with similar standardized recipes, and word them alike for similar procedures.

Selecting a Yield for Testing

To be standardized, a recipe must be prepared in the quantity that is usually served. It is generally necessary to prepare the recipe several times to complete the standardization. The quantity prepared the first time will vary according to the yield of the original recipe because a recipe usually is increased from the original size by doubling the quantity, preparing the recipe, and evaluating it. When a satisfactory product is made in this quantity, it is ready to be increased again. This allows the recipe to be evaluated for quality and use of quantity equipment while controlling the cost of the standardization process.

If the original recipe was a tested quantity recipe, half the number of portions is used as a starting point. If ingredient or equipment substitutions are necessary, it is better to make only one-quarter of the usual production quantity. Recipe adjustments are explained later in this chapter.

Planning for Evaluation

Before the recipe is tested, some thought should go into how it will be evaluated. Who should taste the product? How will their comments be used in evaluating the recipe? Whenever possible both food production workers and clients should help taste and evaluate new recipes. A simple form can be made to record everyone's opinion. An example of one type of evaluation form is shown on page 14. It can be changed by adding questions specific to the food being tasted.

Planning for Production

Preparing recipes for standardization will take longer than preparing a recipe that is already standardized. The recipe will be new to the cook, and extra care must be taken in weighing and measuring during the testing. If ingredients are not on hand, they should be added to the regular purchase order. Testing should be scheduled for a day when there will be enough time. When larger batches are being tested, consideration should be given to how the food can be used. Facilities may serve the product in the employee cafeteria, as an alternative choice to clients, or as an addition to the menu.

Preparing the Recipe

When the recipe is being prepared, each ingredient should be double-checked. Making sure the correct item is being used in the specified quantity is very important. Any changes from the written recipe should be written down during preparation.

Times should be recorded for steps that were timed, such as mixing, cooking, and cooling, as well as the total time. The exact equipment used is listed as part of the recipe directions. For mixers, record the bowl size, type of paddle or whip, and mixing speed. For cooking, indicate if the pan was covered or uncovered.

SELECTING THE CORRECT PAN. Pans come in many sizes. To have a good product, the correct pan must be used. A pan that is too large may cause the product to cook too fast and lose a lot of water in evaporation. If a pan is too small, cooking may take longer, the center may not get done, or the product may cook over the sides of the pan. Table 1.5 lists the most common pan sizes for quantity food production. The correct pan can be selected by comparing the yield in volume or portions with this table. A pan yield in portions can vary, depending on the dipper size or cutting method (Figure 1.1). Table 1.5 can be used to determine how many portions a pan will yield when a specific portion size will be served.

Recipe Evaluation Project

Recipe Name: _____   Name: _____

                                 Today's Date: _____

Directions: Evaluate the food's appearance, taste, texture, and temperature when you get your food sample.

1. Notice the outside appearance and aroma. Does it look and smell appetizing?

2. Taste the food. How does it taste? Is the temperature right? How does the food feel in your mouth? Does it leave a good aftertaste?

3. If you have any other suggestions or comments, write them on the back of the form or tell the foodservice worker in charge of the tasting session.

Answer the questions listed below by circling the word that is closest to your evaluation of this food.

How does the food **smell**?	Great	Good	OK	Bad
How does it **taste**?	Great	Good	OK	Bad
How is its **temperature**?	Great	Good	OK	Bad
How does it **look**?	Great	Good	OK	Bad
How does it **feel in your mouth**?	Great	Good	OK	Bad
How is the **aftertaste**?	Great	Good	OK	Bad
How is it **overall**?	Great	Good	OK	Bad

How **many times a month** should it be served?

   5   4   3   2   1   0

Write any comments of suggestions about the food on the back of the form.

Thank you for helping us with this taste test.

Portion Control

An important part of recipe standardization is selecting the portion size. The yield of the recipe is based on portions that are all the same size. The yield of the recipe being tested must be determined. To accurately count the yield in portions, extra care must be taken to make every portion the correct size.

The portion size is decided after thinking about such factors as

- Traditional portion size
- Nutritional needs of the client

- Appearance of the portion
- What else will be served with the product
- Cost
- Appetite of the client
- Size of the china used for the item
- Policies of the facility, government, or accrediting agencies

Table 1.5. Pan dimensions and capacity in volume and portions

Dimensions	Capacity	Products Measured with Dippers			
		Portion Size	Ladle Size	Dipper Number	Number of Portions
12×20×2½"	2 gal	½ c	4 fl oz	8	64
		⅜ c	3 fl oz	10[a]	85
		⅓ c		12	96
		¼ c	2 fl oz	16	128
12×20×4"	3½ gal	½ c	4 fl oz	8	112
		⅜ c	3 fl oz	10[a]	149
		⅓ c		12	168
		¼ c	2 fl oz	16	224
12×20×6"	5 gal	½ c	4 fl oz	8	160
		⅜ c	3 fl oz	10[a]	213
		⅓ c		12	240
		¼ c	2 fl oz	16	320

Dimensions	Products Cut into Servings		
	Portion Size	Pan Cutting Directions	Number of Portions
12×20×2½"	3×3⅓"	4×6	24
	2½×4"	5×5	25
	3×2½"	4×8	32
	2×2½"	6×8	48
18×26×1"	3½×5¼"	5×5	25
	4½×3¼"	4×8	32
	6×3¼×6½"	3×8×2[b]	48 (triangles)
	3½×2½"	5×10	50
	3×3¼×4½"	6×8×2[b]	96 (triangles)
	1¾×2½"	10×10	100

[a]A level no. 10 dipper is actually 3.2 fl oz.
[b]Cut diagonally.

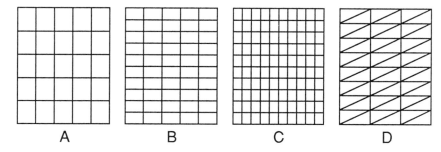

Figure 1.1. Cutting diagrams. A. 5×5, 25 portions; B. 5×10, 50 portions; C. 10×10, 100 portions; D. 3×8, then cut diagonally, 48 portions.

Many tools are available to assist in portioning foods so that all portions are the same size. For products that must be sliced, the most uniform slice thickness will be obtained by using an electric meat slicer. When a slicer is not available or appropriate, uniform slices can be made by using a sharp knife and a washable plastic tape measure or ruler or a pan marker. The food type and size will determine the appropriate knife to use.

The utensils usually used to portion soft or liquid foods are illustrated in Figure 1.2. Liquid and soft foods like soups, sauces, puddings, and some casseroles can be served using ladles, dippers, or spoodles. These utensils come in a variety of standard sizes. Because the food can be leveled in the utensil and the tool has a known capacity, portions can be served easily and quickly and still all be the same size. Use Table 1.6 as a reference when describing portion sizes and serving directions.

SLICING EQUIPMENT. Many types of electric and manual tools are available for slicing foods. Electric slicers, usually called meat slicers, can be easily adjusted to vary slice thickness. To make portions of roast meat all the same weight, the slice thickness can be changed when the diameter of the roast increases or decreases.

The knives most commonly used in quantity production are illustrated in Figure 1.3. The name of the knife indicates its primary use. It is important to keep knives sharp. Using a dull knife increases the loss of product from poorly sliced portions, causes the meat to lose more of its juices, eliminates the possibility of thin slices, and increases the risk of accidents. The sharpening steel shown in Figure 1.4 is used to resharpen knives. Following these diagrams, knives of all types can be easily resharpened. The resharpening smooths the edge of the blade, removing small irregularities. Large nicks or dents require more extensive repair. Using the steel correctly will bring the knife blade to a sharp, smooth edge. The resharpening process, called *steeling* the knife, also realigns the components of the metal, which helps keep the knife sharp. To maintain a sharp edge on all knives, they should never be used on metal surfaces, such as metal counters or pans. Use a dull metal table knife or dough block to cut portions in pans to prevent dulling the sharp knives and to prevent scratching the pans.

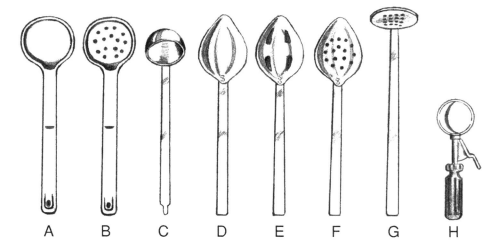

Figure 1.2. Utensils commonly used for portioning soft and liquid foods. A. Spoodle. B. Perforated spoodle. C. Ladle. D. Spoon. E. Slotted spoon. F. Perforated spoon. G. Perforated skimmer. H. Dipper.

RECIPE STANDARDIZATION

Table 1.6. Common dipper and ladle sizes, capacities, and uses

	Dippers Approximate Capacity		
Dipper Number	Weight	Measure	Use
6	6 oz	¾ c	Soups
8	4–5 oz	½ c	Luncheon entrees, potatoes
10	3–4 oz	⅜ c	Desserts, meat patties, ice cream
12	2½–3 oz	⅓ c	Vegetables, desserts, puddings
16	2–2¼ oz	¼ c	Muffins, cottage cheese, croquettes, desserts
20	1¾–2 oz	3 TBSP ¾ tsp	Muffins, cupcakes, meat salads
24	1½–1¾ oz	2 TBSP 2 tsp	Cream puffs, ice cream
30	1–1½ oz	2 TBSP ¾ tsp	Drop cookies
40	¾ oz	1 TBSP 2¼ tsp	Whipped cream, toppings, gravy
60	½ oz	1 TBSP	Salad dressings, toppings
70	⅓ oz	2¾ tsp	Cream cheese, salad dressing, jelly
100	¼ oz	2 tsp	Whipped butter

	Ladles Approximate Capacity		
Ladle Size	Weight	Measure	Use
1 oz	1–1¼ oz	2 TBSP	Sauces, salad dressings, cream
2 oz	1¾–2¼ oz	¼ c	Gravies, sauces
3 oz	2¾–3¼ oz	⅓ c	Cereals, casseroles, meat sauces
4 oz	3½–4½ oz	½ c	Puddings, creamed vegetables
6 oz	5–6½ oz	¾ c	Stews, creamed entrees, soup
8 oz	6–8½ oz	1 c	Soup

PAN MARKERS. For cutting solid foods into portions, marking tools are useful. Pie markers are available to make 6, 7, 8, or 9 equal portions. Expandable dough and cake markers, actually named rolling markers, mark (but do not cut) various size pans and foods into 6 or 8 rows. A washable plastic tape measure or ruler also can be used to divide pans into equal portions.

DIPPERS. Dippers are identified by number. The number means how many level dippers can be obtained from 1 quart. The number is engraved on the curved scraper blade inside the bowl that ejects the food (see Figure 1.2). Some dippers also have a color-coded dot on the end of the handle. Dippers can be purchased for use with the right hand, left hand, or for either hand. When using dippers, care must be taken to prevent all the food on a plate from having the same shape. With practice, a dipper can be used without the food looking like "scoops of ice cream."

LADLES. Ladles are sized by their fluid ounce capacity. This volume measurement should not be confused with the weight of the portion. A 4-ounce ladle will serve 4 fluid ounces, which equals ½ cup. Whether this portion weighs 4 ounces depends on what product is being served. A 4-ounce ladle of spaghetti sauce with meat will weigh approximately 5 ounces and a 4-ounce ladle of fresh strawberry slices will weigh approximately 2.5 ounces.

SPOODLES. Spoodles, a combination of a ladle and a spoon, are illustrated in Figure 1.2. They have a standard capacity and are used for serving semisolid foods and vegetables. Like spoons, spoodles are also available with perforations in the bowl to allow extra juices to drain.

SPOONS. Solid and perforated spoons are frequently used for serving food. It is very hard to make every portion the same size when using a spoon. Weighing a portion at the beginning of service and frequently during service gives better portion control.

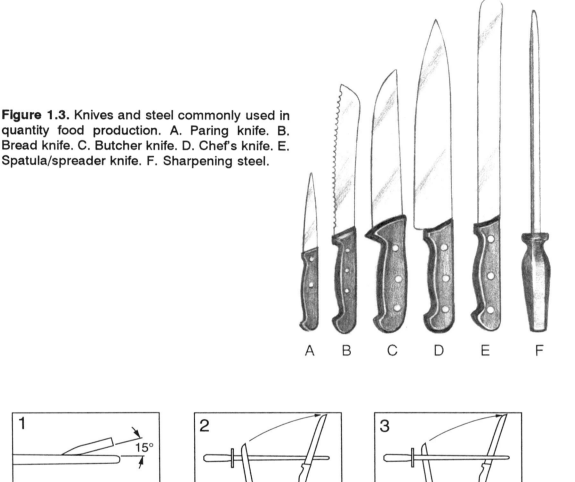

Figure 1.3. Knives and steel commonly used in quantity food production. A. Paring knife. B. Bread knife. C. Butcher knife. D. Chef's knife. E. Spatula/spreader knife. F. Sharpening steel.

Figure 1.4. Resharpening a knife. 1. Place the knife blade on the steel at a 15° angle, as shown. 2. Hone the knife with light arching strokes against the steel. 3. Stroke the knife on alternating sides of the steel to hone both sides of the knife.

A sample portion that is carefully weighed or measured at the beginning of service can be wrapped and placed at the serving line as a standard for the rest of the portions during service.

COMPLETING THE STANDARDIZATION OF A RECIPE

The next step in the process of standardizing the recipe is to record the total yield in weight and/or volume and total number of pans and pan sizes. These should be added to the recipe header along with the cooking time and temperature.

Evaluating the Recipe

The person responsible for recipe testing should review information from several sources to help decide if the recipe needs further testing or is ready to be used.

- Evaluation forms from the taste panel
- Comments from the cook
- Cost
- Yield information
- Availability of ingredients
- Amount of labor required

If a recipe is not yet standardized, it should be scheduled to be tested again, with changes suggested by the information collected. If the recipe was mostly successful, a larger yield can be prepared the next time. If there were problems with quality and quantity, the same-size batch should be made again.

Identifying Solutions to Quality Problems

To make any necessary changes to improve product quality, the factors causing the problem must be identified. This study course has information about ingredients that will help identify the problems. Depending on knowledge of how food reacts under different conditions is better than using trial and error.

Final Testing of the Recipe

When solutions have been found, a second recipe-testing session should be scheduled. The process may be repeated a third and fourth time, if necessary. Most recipes must be prepared at least three times to develop a good standardized recipe.

The final test of the recipe is having it prepared by a cook who has not used it before. A standardized recipe should produce the same good quality and equal quantity when prepared by any trained cook in a facility. When the recipe passes this test it is ready to be planned into the menu cycle.

In health care facilities a recipe is usually analyzed for the nutrient content. When nutritional information is important, this analysis should be done before the recipe is added to the menu.

ADJUSTING RECIPES

Adjusting recipes to the exact quantity needed will prevent having leftovers or running out of food. In these nursing facilities where the number to be served does not often change, daily adjustments may not be necessary. But if costs are to be controlled, and client satisfaction is important, adjustments will be a frequent part of production planning.

Adjustment Methods

There are several methods to adjust recipes. The factor method is recommended for recipe adjustments because it is accurate, can be used to adjust to any yield needed, and is relatively easy to learn. Others include direct-reading tables and the rarely used percent method.

FACTOR METHOD. The factor method is most easily completed when it is separated into a sequence of steps. The same steps are followed for increasing or decreasing a recipe. Since each ingredient quantity will be multiplied, a calculator is helpful. Because of the calculations, it is easier to use pounds and decimal parts of a pound than pounds and ounces. A conversion table of decimal parts of a pound, such as Table 1.3, is helpful.

Step 1. Divide the yield wanted by the yield of the standardized recipe. The result is the *factor*.

Example 1. The facility needs to serve 112 portions of oven-fried chicken. The standardized recipe is for 200 portions. Divide 112 by 200.
$$112/200 = .56 = \text{the } factor$$

Example 2. 112 portions of cake are needed. The recipe is for 80 portions baked in two 12×20×2½″ pans.
  A. The portions prepared of items baked in pans, such as cakes, pies, and casseroles, must be rounded to whole or half pans before the factor is determined.
$$112/40 \text{ portions per pan} = 2.8 \text{ pans}$$
  B. Identify the number of pans to be produced. Here 2.8 pans are needed, so 3 pans will be produced.
  C. Identify the actual number of portions that will be produced.
$$3 \text{ pans} \times 40 \text{ portions per pan} = 120 \text{ portions}$$
  D. To calculate the factor, divide the portions to be produced by the portions listed in the original recipe.
$$120 \text{ portions}/80 \text{ portions} = 1.5 = \text{the factor}$$

Step 2. Convert all ingredient quantities to easy-to-use units. Use a conversion table, such as Table 1.3, or a calculator. If a calculator is used, convert all recipe ingredient quantities to whole units or decimals rather than fractions. (The quantities will be converted back to pounds and ounces after the adjustments are finished.)

Example 1. The original recipe uses ¾ cup plus 2 tablespoons of salt:
  A. Convert the quantity to whole units or decimal parts of a unit.
    For whole units, ¾ c = 12 TBSP, so 12 TBSP + 2 TBSP = 14 TBSP.
    For decimal parts of a unit, convert ¾ c + 2 TBSP to decimal parts of a cup (14 TBSP is what decimal part of a cup?). Divide the recipe quantity by the number of units in the whole unit. In this example the correct formula is 14 TBSP/16 TBSP because there are 16 TBSP in a cup.
$$14/16 = .875 \text{ c}$$

Example 2. 4 lb 7 oz flour is used in the original recipe. Using Table 1.3 is the easiest way to convert this quantity to pounds and decimal parts of a pound.
  A. 4 lb 7 oz = 4 lb + the decimal pound equivalent of 7 oz.
  B. Using Table 1.3, 7 oz = .438 lb. Therefore 4 lb 7 oz = 4.438 lb.

Example 3. Convert the 4 lb 7 oz of flour to whole, smaller units, such as ounces.
$$4 \text{ lb } 7 \text{ oz} = (4 \times 16 \text{ oz}) + 7 \text{ oz} = 64 \text{ oz} + 7 \text{ oz} = 71 \text{ oz}$$

Step 3. Multiply the *factor* found in Step 1 by the quantity of each ingredient identified in Step 2.

Example 1. Recipe original yield is 200 portions, needed yield is 112 portions, *factor* = 112/200 = .56.

Ingredient	Original Amount	× .56	Adjusted Amount
Chicken	200 quarters	× .56	112 quarters
Flour, all purpose	4 lb 7 oz	× .56	2 lb 7.76 oz
	4.438 lb	× .56	2.485 lb
	or 71 oz	× .56	39.76 oz
Salt	¾ c 2 TBSP	× .56	7 TBSP 2.5 tsp
	or 14 TBSP	× .56	7.84 TBSP
	or .875 c	× .56	.49 c
	or 42 tsp	× .56	23.5 tsp
Pepper, black	or ⅓ c (or .33 c)	× .56	.185 c
	or 5⅓ TBSP	× .56	2.99 TBSP

Step 4. Convert the resulting amounts back into the measuring unit that is most appropriate for the facility and the ingredient quantity, using a conversion table or calculations.

Example 1. Calculation method of conversion.

Ingredient	Amount	Calculation		Converted Amount
Flour	2.485 lb	.485 lb × 16 oz/lb	=	2 lb 7.76 oz
Salt	.49 c	.49 × 16 TBSP/c	=	7.84 TBSP
	or 7.84 TBSP	.84 TBSP × 16/c	=	7 TBSP 2½ tsp
Pepper	.185 c	.185 × 16 TBSP/c	=	2.96 TBSP

Step 5. Round off the ingredient amounts using a rounding table (see Table 1.4).

Ingredient	Amount	Rounded Amount
Flour	2 lb 7.76 oz	2 lb 8 oz
Salt	7 TBSP 2½ tsp	7 TBSP 2½ tsp
Pepper	2.96 TBSP	3 TBSP

Step 6. Write the rounded adjustments on the recipe.

DIRECT-READING TABLE. The direct-reading-table method involves tables that are simple and quick to use. They require little math skill. The disadvantage of the direct-reading table is that recipes can only be adjusted in multiples of 25 portions: 25, 50, 75, up to 500.

PERCENT METHOD OF RECIPE ADJUSTMENT. An accurate but less frequently used adjustment procedure is the percent method. To use this method, first convert all ingredient measurements to weight and calculate the total weight of the recipe. Next calculate the percent of this total represented by each ingredient. Finally, determine the percent required of each ingredient.

Although the percentage method is as accurate as the factor method, it is considered more difficult to use. The conversion to and from percents, weights, and measures is more likely to result in errors.

Adjusting Seasonings. Herbs and spices used in recipes are adjusted according to recipe quantity using the factor method, just like all other ingredients. To retain the same flavor, the seasonings must be increased or decreased by the same factor. Changes in cooking or holding time or cooking equipment may require additional adjustment of seasonings.

When preparing a very small quantity of a recipe standardized in large quantity or a large quantity of a recipe standardized in a much smaller quantity, problems with seasoning may result. The recipe may not be successful unless the size of mixing and cooking equipment is changed to be in proportion to the new quantity. A recipe that usually makes 10 gallons of soup using a 20-gallon steam-jacketed kettle (SJK) can be adjusted to make only 1 gallon. If 1 gallon of soup is cooked in a 20-gallon SJK, too much of the water will evaporate because of the relatively large amount of surface area. This will make the seasonings seem too strong. If the 1 gallon of soup was made in a 5-gallon SJK, the seasonings would be correct.

In another example, 1 gallon of chili requires 1 hour of simmering time in a 5-gallon SJK. When 15 gallons are made in a 20-gallon SJK, the chili must simmer 2 hours to reach the same thickness as the 1 gallon of chili. This extra simmering, not the recipe adjustment, will increase the spiciness of the chili. Usually herbs decrease in strength when cooked longer, but peppers and chilies increase in flavor strength as the amount of cooking and holding time increases.

SUMMARY

Standardized recipes make it possible to serve food of the same quality in a known quantity every day. The recipe standardization process is a way to develop recipes for a facility. Both quality and quantity are checked. Obtaining ideal quality is just as important as consistent quantity. With standardized recipes and accurate recipe adjustments, every menu item should be good and be available in the correct quantity. To adjust recipes, the factor method is recommended. It is better than the direct-reading-table method because the recipe can be adjusted to the exact quantity and is likely to be more accurate to use than the percentage method.

The standardization process consists of 10 steps:

1. Selecting a recipe format
2. Writing the recipe in the selected format
    a. Yield and equipment information in the recipe header
    b. Ingredients and directions in the recipe body
3. Planning for evaluation of the recipe
4. Planning for production
5. Preparing the recipe
    a. Selecting the correct pan
    b. Identifying the portion size and utensil
6. Evaluating the product and recipe
7. Identifying solutions to problems
8. Scheduling the recipe to be produced at least once again
9. Final testing with another worker making the product
10. Putting the recipe on the menu

LEARNING ACTIVITIES

Activity 1: Writing a Recipe

1. Select a recipe used in your facility that is not already standardized or a recipe from a small-quantity cookbook.

2. Rewrite the recipe using the block format.

RECIPE STANDARDIZATION

Activity 2: Adjusting a Recipe

1. Use the factor method of adjustment to adjust the Macaroni and Cheese recipe to 85 portions.

    A. Convert gallons, quarts, cups, and pounds and ounces to decimal units, using Table 1.3.

    B. Convert the amounts back to appropriate whole units, fractions of units, and pounds and ounces.

    C. After adjusting the recipe, round the quantity of each ingredient correctly.

2. Fill in the chart.

Activity 2, Step 2

Macaroni and Cheese			
Yield: 2 pans 12×20×2″ = 48 portions			

Ingredient	Amount for 48 Portions	Amount Needed for 85 Portions	
		In decimals	In fractions
Macaroni, cooked 10 lb 8 oz			
Margarine, melted 12 oz			
Flour, all-purpose 8 oz			
Salt 1 oz			
Dry mustard 1 TBSP			
Worcestershire sauce ¼ c			
Milk, 2% 1 gal			
Cheese, cheddar, grated 4 lb			
Bread crumbs 1 lb 4 oz			
Margarine, melted 7 oz			

REVIEW QUESTIONS

True or False

1. The block recipe format is recommended for quantity food production organizations.

    A. True
    B. False

2. Standardized recipes are available for purchase from governmental agencies.

    A. True
    B. False

3. In a recipe with ingredients that are weighed, the adjusted amounts should be rounded to the nearest whole ounce.

    A. True
    B. False

4. Most recipes can be standardized by being carefully prepared once.

    A. True
    B. False

Multiple Choice

5. The size of a dipper is equal to the number of

    A. Ounces it holds
    B. Level tablespoons it holds
    C. Level dippers in a pound
    D. Level dippers in a quart
    E. Level dippers in a gallon

6. The direct-reading-table method of adjustment will help to

    A. Adjust a recipe to the exact quantity needed
    B. Round recipe adjustments to whole pans
    C. Adjust a recipe to 125 portions
    D. Adjust a recipe to 96 portions

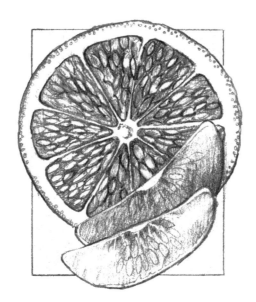

2. FRUITS

Fruits are popular at all meals as well as for snacks. They can be served raw, whole or sliced, in salads or as desserts or snacks. They add contrast in texture, flavor, and color. Fruits are an important source of vitamins and minerals. Yellow fruits like peaches, cantaloupes, and apricots are excellent sources of vitamin A due to their high level of carotene. Citrus fruits are the main sources of vitamin C in the diet. Bananas provide potassium, and dried fruits, such as raisins, are a source of iron. Some raw fruits also are an excellent source of fiber when the peel is eaten.

UNDERSTANDING FRUITS

Classification

Fruits are defined as the edible part of a plant or tree, consisting of the seeds and pulpy surrounding tissue. Usually they are distinguished from vegetables in their use in the main course of a meal. Fruits are classified by their shape, structure, type of seed, and where they are grown.

Classification of Fruits	Examples
Berries	Strawberry, cranberry, grape, raspberry, blueberry, currant
Citrus fruits	Orange, lime, lemon, grapefruit, tangerine
Drupes (stonefruits)	Cherry, plum, peach, nectarine, apricot, prune
Melons	Cantaloupe, watermelon, honeydew, casaba
Tropical fruits	Avocado, banana, date, kiwi, pineapple
Pomes	Apple, pear

Structure and Composition

Fruits and vegetables consist of four parts: an outer covering, a vascular system, edible flesh, and supporting tissues. Most fruits have a high water content that provides a firm, juicy texture.

Carbohydrates are a main component of fruits. The type of carbohydrate in fruits varies with the maturity of the fruit, but the total amount of carbohydrate remains almost constant. Starch in immature fruits changes into sugar as they ripen. This can be detected in the sweet flavor of ripened fruits.

Pectic substances, the complex carbohydrates in fruit, help to give fruit its texture. The change in texture of fruits as they ripen is due to changes in pectic substances. Enzymes in fruits are responsible for the ripening process, including the development of sugar and aroma. Enzyme reactions also cause the browning on the cut or peeled surface of some fruits.

PIGMENTS. The attractive colors of fruits are due to the pigments they contain. In addition, they are also responsible for some color changes during cooking and as fruits ripen.

Pigments	Colors	Fruits
Chlorophyll	Green	Avocado, greengage plum, gooseberry
Carotinoid	Orange	Apricot, muskmelon, orange, peach, pineapple, tomato, red grapefruit, watermelon
Anthocyanin	Red, blue	Strawberry, blueberry, sweet cherry, concord grape
Anthoxanthin	White	White membrane of orange

These pigments are also found in vegetables, and their specific characteristics and interactions with acid and alkaline are discussed in Chapter 3.

Generally, color changes during cooking are not a problem with fruit. When fruits containing anthocyanin, such as concord grapes, cherries, and black raspberries, are combined with an acidic juice like orange or lemon juice, the color of the mixture will be acceptable. However, these fruit juices may turn blue if the water is hard or alkaline. Problems may develop when these fruits are added to quick breads. The alkaline leavening agents react with the pigment, resulting in a blue or green color. If a large amount of orange juice is added to red or purple fruit juice, an unappealing brownish color is produced.

The pigments in strawberries and cranberries are quite stable, and they are used as ingredients when making red-colored fruit juice mixtures. Both anthocyanin and anthoxanthin change in the presence of metal ions, such as aluminum, iron, or tin. Therefore, fruit knives and pans made from aluminum or iron should be avoided.

SELECTION

Fruits are available in fresh, canned, frozen, and dried forms. Each of these has certain advantages. When purchasing fruits in a particular form, their use, storage, and nutritional value must be considered.

Fresh Fruits

Fresh fruits are graded by the U.S. Department of Agriculture (USDA). The grades are designated by number or name. In quantity foodservices, the two most common grades used are

- U.S. Fancy: the highest quality available
- U.S. No. 1: most commonly available produce

The quality of fresh fruit depends on its maturity and ripeness. *Maturity* means fruit has reached its full size. *Ripeness* refers to how ready it is to eat.

Fresh fruit is purchased by weight or count rather than size. However, the size is reflected in the count. For example, the larger the size, the fewer in a case and therefore the lower the count. Medium size is usually more desirable than extra large or very small sizes. Following is a purchasing guide for fresh fruits.

Fruit	What to Look For	What to Avoid
Apple	Firm, crisp; good color for variety	Overripe, soft or mealy, bruised
Apricot	Uniform, golden color; plump, juicy; barely soft	Soft or mushy; hard; yellow or green color
Avocado	Firm if to be used later, slightly soft for immediate use	Dark patches; cracked surfaces
Banana	Firm, bright color; no bruises	Bruised; discolored skin
Blueberry	Dark blue with silver bloom; firm, plump; uniform size	Soft, spoiled berries; stems and leaves
Cherry	Dark color in sweet cherries, bright red in pie cherries; glossy, plump	Shriveling; soft; leaking fruit; dull appearance; moldy
Cranberry	Plump and firm; lustrous red color	Soft and spongy; leaking
Grapefruit	Firm; well-shaped; heavy for size; thin-skinned; juicy	Soft and discolored areas; moldy
Lemon	Rich yellow color; firm; heavy	Hard or shriveled; soft spots; moldy; dark yellow
Lime	Glossy skin; heavy	Dry skin; decayed
Melons		
Cantaloupe	Smooth area where stem grew; bold netting; yellowish cast to skin	Soft rind; bruised; moldy
Honeydew	Faint odor; yellow to creamy rind; slight softening at blossom end	Greenish-white rind; hard and smooth skin
Watermelon	Slightly dull rind; creamy color on the underside	Cracks; dull rind
Orange	Firm and heavy; bright, fresh skin with either orange or green tint	Light weight; dull skin; moldy
Peach	Slightly soft; yellow color between the red areas	Very firm, hard; green ground color; very soft; decayed
Pear	Firm, but beginning to soften; good color for variety (Bartlett, yellow; Anjou or Comice, light green to yellow green; Bosc, greenish yellow with skin russeting; Winter Nellis, medium to light green)	Weakening around the stem; spots; shriveled
Pineapple	Good aroma; green to yellow color; spike leaves easily removed; heavy for size	Bruised; poor odor; sunken or slightly pointed pips; soft
Plum	Fairly firm; good color for variety	Hard or shriveled; poor color; leaking
Strawberry	Good red color; clean; lustrous; cap stem attached	Moldy; leaking; large seeds

Processed Fruits

In quantity foodservice operations, canned and frozen fruits are often served, instead of fresh fruits. This decreases labor and high food costs associated with the cleaning, handling, and preparing and possible spoilage of fresh fruit. For example, the individual sectioning of citrus fruits could be too expensive for some facilities. Therefore, a no. 10 can or 1-gallon jar of citrus sections might be used. Some canned or frozen products will be close to the flavor and quality of fresh fruit, although they are usually softer in texture.

FROZEN FRUITS. Frozen fruits may be processed with or without sugar. Grade A or Fancy products are used for freezing fruits. Frozen fruit usually has the same bright color and flavor of fresh fruit. The texture tends to be softer, because the ice crystals break the fruit cell walls during the freezing process.

For some recipes, frozen fruit should be thawed in the refrigerator in the unopened package and then used immediately. If only the pieces of fruit are to be used in the recipe, the thawed fruit must drained before it is used. Otherwise, there will be too much liquid in the recipe.

When frozen berries and melons are served, they should be portioned while frozen solid if individually quick-frozen and served when they are partially thawed. Blueberries are added while still frozen in most quick-bread recipes.

CANNED FRUITS. Canned fruit can be served by itself or as an ingredient in salad or dessert. The finest grade, U.S. Grade A, should be used when the fruit is served alone. U.S. Grade B can be used in making fruit mixtures, like fruit pies, where perfectly uniform pieces are not needed. Canned fruits come packed with various liquids and in several solid forms. If fruit is packed in solid form without added juice, it is called *pie pack*. Fruit comes packed in heavy syrup, light syrup, and juice pack. Different forms and syrup sweetnesses are best suited for different recipes. Here are the grading standards for canned fruits.

Grade	Quality of Fruit	Syrup
U.S. Grade Fancy or U.S. Grade A	Excellent quality; firm; good color; ripe; free from blemish; uniform in size; very symmetrical	Heavy; about 55% (40–70%) sugar, depending on acidity of fruit
U.S. Grade Choice, U.S. Grade B, or Extra Standard	Fine quality, ripe; firm; free from most blemishes; good color; uniform in size; and symmetrical	About 40% sugar, 10–15% less than Fancy Grade
U.S. Grade C or Standard	Good quality; reasonably free from blemishes; reasonably good color; reasonably uniform in size, color, and degree of ripeness; reasonably symmetrical	About 25% sugar, 10–15% less than Choice Grade
Substandard	Lower than minimum grade for Standard	Often water or solid pack; if packed in syrup, not over 10%

DRIED FRUITS. Dried fruits may be eaten as snacks or used for cooking and baking. The most common dried fruits are raisins, prunes, dates, and apricots. Sometimes the drying process makes the products tough. Dried fruit should have good color and a soft texture. Dried prunes are also sold moist pack and in juice.

FRUIT JUICES. Juices labeled fruit juice must be 100% fruit juice without added water. Juices mixed with water must be labeled *fruit drink*. Fruit juices are available in fresh, frozen, canned, or dehydrated crystals.

One of the most popular juices is orange juice, and all market forms of orange juice contain about the same amount of vitamin C. Canned juices are fortified to maintain the same vitamin C content as fresh orange juice. Once the canned orange juice is opened, the remaining portion should be transferred to nonmetal containers and stored in the refrigerator. It should also be covered to prevent air from reducing the vitamin C content.

Grape, apple, prune, pineapple, and cranberry juices and apricot nectar contain little or no vitamin C. For this reason, unless they are fortified they cannot be substituted for citrus juice as a source of the vitamin C.

STORAGE

Fresh Fruits

Fruits are usually picked when they are still unripe. They are often produced far from the markets where they will be purchased by consumers, and the damage that may occur during shipping is reduced when fruit is unripened. Bananas and avocados are harvested when they are still firm and unripe and continue to ripen during shipping and storage. These fruits should be stored at room temperature until they are ripe. Other fruits should be refrigerated to maintain flavor and firmness. Apples and citrus fruits can be stored for a long time, but most fruits must be used within a few days.

The keeping quality of fruit depends on how fresh it was when it was purchased, how it was handled, and the storage conditions. Use these guidelines for storing fresh fruits.

Fruit	Storage Conditions
Underripe fruit	Allow to stand at room temperature to ripen. If desired, put fruits in a brown paper bag for faster ripening.
Banana	Ripen at room temperature in an open container. Refrigeration after ripening will cause dark skin, but it preserves the quality of the fruit for use in baked products.
Citrus fruit	Store at room temperature or refrigerate for longer storage.
Melon, pineapple	Refrigerate. Wrap in food film to keep the aroma from flavoring other foods.
Berry, cherry	Sort to remove damaged and decayed berries. Refrigerate in a closed plastic bag or shallow covered container. Use as soon as possible. Wash just before serving.
Cut fruit	Refrigerate in a stainless steel or plastic covered container. To avoid browning, coat with citrus juice or antioxidants before storing.
All other ripe fruit	Wash before storing. Dry well. Refrigerate loosely wrapped.

Processed Fruits

CANNED FRUITS. Canned fruit should be stored in a cool, dry area that is no warmer than 70 F. The color, flavor, and texture of canned fruit may change if it is stored in a very warm place (above 70 F) or held longer than a year. Because it is not recommended to store either canned or frozen fruit longer than one year, care should be taken to date and rotate stock when it arrives.

Opened canned fruit should not be stored in its original metal container. When exposed to the air, the container will react with the acid in the juice, causing a dark color. Any unused fruit should be transferred to a National Sanitation Foundation (NSF)–approved covered container and refrigerated at 36–40 F.

DRIED FRUITS. Dried fruits are stable if held in a cool, dry environment. Although they can be kept for 6 months or longer, refrigeration may be necessary during hot, humid summer months. Management and storeroom personnel should not have more than a few months' supply at one time.

If dried fruits, such as apples, apricots, peaches, and pears, are stored properly, they can retain their color. The concentration of sugars in dried fruits is high, so humid air should be avoided in the storage area. High humidity will cause sugar to crystallize on the surface, causing a dull, unappetizing appearance.

PREPARATION

In order to maximize the quality and nutrients found in fruits, quantity foodservice operations should choose preparation methods that will retain texture, color, flavor, and nutrients. To retain their best quality, preparation should be scheduled as near the time of service as possible. Minimum handling of fruit is recommended because overhandling causes bruises and may cause the fruit to spoil.

Pieces of fresh fruit should be large enough to be identified. They are more attractive and retain nutrient value better than a finely chopped product. When fruits are strained or mashed, the loss of vitamin C will increase.

Fresh Fruits

WASHING. All fruits should be washed to remove surface soil, sprays, and preservatives before serving or cooking. Usually apricots, cherries, grapes, pears, and plums are washed in a colander under cold running water. Some fruits, such as apples and plums, may be washed and dried before being refrigerated. The tough skins protect their flesh from bruising during washing.

Berries deteriorate rapidly, so washing should be done right before service. They need to be handled gently during preparation. Stems of strawberries should be removed after washing. If stems are removed before washing, the rate of vitamin C loss and bruising will increase and the berries will become too soft. After washing, strawberries should be drained thoroughly and refrigerated.

Citrus fruits are usually served in one of two ways, prepared in sections or cut in half horizontally. A sharp, stainless steel knife should be used in peeling and sectioning because other metals react with acid. Figure 2.1 shows how to peel and serve citrus fruits in sections.

When a citrus fruit is cut in half, the sections should be loosened with a knife. To cut a citrus fruit perpendicularly, peel and then slice perpendicularly to center core.

CONTROLLING QUALITY CHANGE. Many fruits, such as bananas, apples, peaches, and pears darken rapidly when cut as a result of the oxidizing action of the enzymes when they come in contact with air. This is called *enzymatic browning*. Because this change in color is the result of contact with air, the cut surface of the fruit must be protected from the air. Applying a small amount of sugar, citrus fruit juice, or antioxidants can help prevent this browning. The sugar draws water from the fruit to the surface, forming a sugar solution that blocks the air. The vitamin C in citrus fruit juice interacts with the oxygen to protect the natural color of the exposed fruit surfaces. The juice also acts to coat the fruit and keep the air out.

Antioxidant powders can be purchased to prevent the oxidizing action. *Sulfite types should not*

Figure 2.1. Peeling and sectioning citrus fruit. A. Cut off a thick slice at the top and bottom and place the fruit on a cutting board. B. Cut skin from the top to bottom deep enough to remove all white membrane. C. Cut a section free from its membrane on one side. D. Turn the knife and force the blade along the membrane of the next section so that the section falls out. E. Continue sectioning.

be used because they can cause severe allergic reactions and death. Ascorbic acid products are acceptable for nursing facilities. The directions for use should be followed carefully. Overuse will cause the fruit to be too soft and taste bitter, but underuse will not prevent browning.

If fresh fruits will not be consumed before they overripen, they can be cooked or frozen. Freezing and cooking will preserve the fruits for later use. Bananas can be peeled and frozen for future use in baking.

Processed Fruits

DRIED FRUIT. Cleaning is the first step in dried fruit preparation. Dirt should be removed by washing. After washing the dried fruit, it should be soaked in water to *rehydrate* (to reabsorb some of the moisture lost during drying). Soaking in water too long will give a poor, watery flavor rather than improved quality. To soften dried fruits, after soaking in water they may be boiled or soaked in hot water for a short time and then simmered. The amount of soaking and cooking time depends on the size of the pieces of fruit and the amount of cut surface exposed. Boiling reduces the rehydration and cooking time but may result in lower product quality.

Due to the high sugar content of dried fruits, additional sugar is usually not necessary. If added sugar is desired, a *small* amount can be added at the end of the cooking period. Too much sugar can cause the fruit to lose its firm texture. When dried fruits are properly cooked, they will be tender and plump and have good flavor and color, and the flesh will separate easily from the pit.

CANNED FRUITS. Canned fruits are generally used for fruit pies and fruit cobblers. They also are used for shortcakes or are served alone. When canned fruits are served alone, they should be chilled thoroughly.

Before canned fruits are opened, the top of the can should be cleaned with a wet towel. For the best product, follow the draining instructions in the recipe carefully.

Cooking Methods

Many fruits, especially firm or slightly underripe ones, can be successfully cooked. They should be cooked for a short time at a low temperature so they retain their shape, flavor, and color. Fruit can be cooked by poaching, baking, simmering, broiling, or glazing.

POACHING. Whenever a fruit is *poached* (covered with liquid and heated just below the boiling point), the fruit is softened and enzymatic browning stops. Flavors can be preserved by limiting the amount of water used, by controlling heat, and by covering fruit throughout the cooking process.

Just enough sugar needs to be added to fresh fruit to prevent it from absorbing excess water. Without this sugar, buildup of water in the cells would cause them to break, resulting in a soft or mushy product. In the preparation of sauces and pureed fruits, however, this quality is desired, so sugar is added only near the end of cooking to sweeten the product yet allow the fruit cells to break to produce the pureed texture.

BAKING. Fruits like apples and pears can be baked without the removal of their skins. However, to avoid breaking of the skin from built-up steam pressure inside the fruit, it is necessary to partially peel or make a thin cut into the peel around the center of the fruit to allow for expansion and the release of steam.

Perhaps the most popular baked fruits are apples. Usually they are baked with the skin on and the cores removed. The core cavity is often filled with sugar, cinnamon, and butter or margarine. Apple varieties differ in cooking quality. Here are the best uses for each variety.

Variety of Apple	Description	Use
Golden Delicious	Whitish-golden, tender skin, sweet flavor	Raw, cooking, baking
Jonathan	Dark red with tiny white spots, tart flavor	Raw, cooking, baking
Rome Beauty	Bright red, juicy, mild acidic flavor	Cooking, baking
Red Delicious	Elongated shape, 5 knoblike points on bottom, sweet, deep red, lacks flavor	Raw; not good for cooking because it stays very firm

Fruits are used in many kinds of baked products. Berries, dates, raisins, prunes, and figs add flavor to muffins, quick breads, cakes, and cookies. Typically, apples, cherries, peaches, and blueberries are made into pies, crisps, and cobblers, with a sweetened and thickened filling.

SIMMERING. Fruits like apples, peaches, and pears can be cooked by simmering. However, dried fruits are the most commonly simmered, in order to rehydrate and tenderize them. Dried fruits need to be soaked before simmering. Because some of the sugar and water-soluble nutrients are dissolved in the water during soaking, this water can be used to advantage in cooking. Dried fruits usually are simmered until they can be cut easily.

BROILING. Grapefruit, bananas, peaches, and pineapples lend themselves to broiling. However, it is important the broiling be only for a brief period to avoid dehydration and overbrowning. Broiled fruits are an excellent low-sugar, low-calorie alternative to pastry desserts.

GLAZING. Glazing is a cooking method in which fruits are cooked in a heavy syrup made of water and sugar. When fruits are cooked this way, they retain their shape. This is a sweet dessert alternative when the pastry in pies and cobblers is not desired in a diet.

MICROWAVE OVEN COOKING. Small amounts of fruit can be cooked in the microwave oven. Cooking time should be carefully controlled to get an optimal product with good texture. Covering the fruit during cooking allows a more even heat distribution. (Specific microwave cooking procedures in the recipe should be followed.) More information on microwave cooking can be found in Chapter 11.

SUMMARY

Fruits are an important source of vitamins and minerals. They may be classified by their shape, structure, type of seed, or where they are grown. They are available in fresh, frozen, canned, and dried forms. Many fresh fruits spoil quickly and must be used shortly after purchase.

Fruits should be stored properly to retain their color, flavor, texture, and nutrients. Once canned fruits or canned juices are opened, any leftovers should be removed from the can, covered, and refrigerated to retain the vitamin C level.

The preparation methods for fruits should be selected to retain their texture, color, and nutrients. Overhandling usually causes bruises and spoilage. Color changes during cooking are not a problem in fruit cookery. Using processed fruits can save cooking time and labor.

LEARNING ACTIVITIES

Activity 1: Comparing Baked Apples and Glazed Apples

1. Prepare baked apples according to the recipe.

   A. Adjust the recipe for 10 portions, filling in the blanks in the chart.

   B. Prepare 10 portions according to recipe.

   C. Record the cooking time and description of the product on the comparison chart. Select from the words that best describe the baked apples, or use your own words.

2. Prepare glazed apples according to the instructions below.

   A. Fill in the ingredients chart to adjust the recipe for 10 portions.

   B. Wash and core the apples.

   C. Cut a slit in the skin, at a right angle to the core, around the middle of each apple.

   D. Place apples in a flat saucepan with a close-fitting cover.

   E. Fill the center of each apple with a mixture of 1 tablespoon sugar, ⅛ teaspoon cinnamon, and 1 teaspoon butter.

Baked Apples			
Yield: 100 servings Portion size: 1 apple with syrup		Baking temperature: 350 F Baking time: 50 min to 1 hr 15 min	
Amount	Ingredients	Amount	Procedure
....	Apples,* baking, medium, 5–6 oz *Use varieties such as Rome Beauty, Winesap	100	1. Wash apples and remove cores. 2. Place cored apples in baking pans.
....	Sugar, granulated.................. Salt................................... Cinnamon, ground...............	4 lb 2 tsp 2 tsp	3. Combine sugar, salt, and cinnamon; blend. 4. Fill cavities of apples with sugar mixture, approx 1½ TBSP per apple.
....	Water, hot, per pan, approx	2 c	5. Pour hot water in pans. DO NOT POUR OVER APPLES. 6. Bake apples at 350 F for 50 min to 1 hr 15 min. 7. Baste apples with liquid in pan during latter part of baking period. 8. Test for doneness by piercing the interior with a fork. Time will vary with the variety of apple. 9. Serve warm with light cream or Ginger Hard Sauce.

Source: *Standardized Quantity Recipe File* (1971).

Activity 1, Step 1A

Ingredient	100 Portions	10 Portions
Apples, baking, medium	100	
Sugar, granulated	4 lb	
Salt	2 tsp	
Cinnamon, ground	2 tsp	
Water, hot, per pan	2 c	

FRUITS

Activity 1, Step 2A

Ingredient	1 Portion	10 Portions
Apples, baking, medium	1	
Sugar, brown	1 TBSP	
Cinnamon, ground	⅛ tsp	
Butter	1 tsp	
Water, hot, per pan	¼ c	

F. Pour ¼ cup of water per apple in the pan. Cover the pan and place it over low heat.

G. Cook until the apples are tender but not mushy (7–15 minutes).

H. Just before removing the apples from heat, remove the cover and turn them over or baste them with liquid to glaze.

I. Record the cooking time and description of the product on the comparison chart.

Activity 1, Steps 1C and 2I

Characteristics of Product	Baked Apples	Glazed Apples
Cooking time		
Exterior appearance (color; shape, plump, shiny, shrunken)		
Interior appearance (firm, tender, hard, soft)		
Flavor (good, sour, too sweet, too strong, tasteless)		

REVIEW QUESTIONS

True or False

1. Fruits are an important source of vitamins, minerals, and proteins.

    A. True
    B. False

2. Cooking breaks down the fiber of fruit and makes it easier to digest.

   A. True
   B. False

3. As a rule, canned fruits are as nutritious as fresh or frozen fruits.

   A. True
   B. False

4. All fresh juices are high in vitamin C.

   A. True
   B. False

5. Starches in immature fruits change into sugar as they ripen, but the total amount of carbohydrates remains relatively the same.

   A. True
   B. False

Multiple Choice

6. Which of the following groups of fruits are the best source of vitamin A?

   A. Peaches, cantaloupes, and apricots
   B. Apples, pears, and watermelons
   C. Berries, bananas, and currants

7. The most common solution to the problem of browning in peeled fruit is the use of

   A. A heavy concentration of sugar
   B. An acidic fruit juice
   C. A mixture of vinegar and water

8. What should you do if you received a shipment of green, unripe bananas?

   A. Place them in refrigerator in their original container until the desired ripeness is achieved.
   B. Keep them sealed inside the plastic bags in which they are shipped.
   C. Open the plastic shipping bags and allow them to ripen at room temperature.

9. Upon receiving a flat of strawberries, what procedure should you follow?

   A. Wash and sort them prior to refrigeration.
   B. Sort and refrigerate them in their natural state.
   C. Wash, remove stems, and sprinkle them with sugar prior to refrigeration.

Matching

10. Match the word related to fruits in column A with the definition in column B by writing the correct letter from column B by each number in column A.

Column A	Column B
C 1. Produce	A. Fruits that are ready to eat
B 2. Mature	B. Fruits that have reached their full size
E 3. Drupes	C. Fresh fruits and vegetables
F 4. Pomes	D. Avocados, bananas, dates, kiwis
G 5. Melons	E. Cherries, plums, peaches, prunes
H 6. Berries	F. Apples, pears
	G. Cantaloupes, watermelons
	H. Strawberries, cranberries, grapes

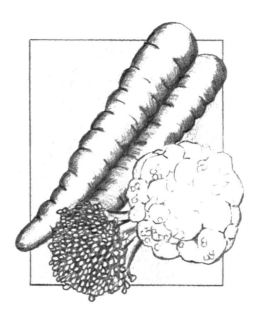

3. VEGETABLES

Vegetables add color, shape, texture, and nutritive value to meals. Modern methods of processing, transporting, and storing vegetables make them available in fresh, frozen, canned, and dried forms all year.

Vegetables, especially fresh ones, are among the best sources of vitamins and minerals. Most yellow and green vegetables are good sources of vitamin A. Green peppers, tomatoes, broccoli, and Brussels sprouts are good sources of vitamin C. *Legumes* (dried seeds from beans and peas) are fairly high in thiamine. Leafy vegetables contain folacin, a B vitamin. Broccoli and spinach are good sources of calcium.

Vegetables, except for starchy vegetables like corn, peas, and potatoes, tend to be low in calories. Starch is the main source of calories in vegetables and the form in which carbohydrate is stored. The protein and fat contents of most vegetables are generally very low. Legumes are an exception, however, with a much higher protein content than other vegetables.

UNDERSTANDING VEGETABLES

Characteristics

STRUCTURE. The structure of vegetables consists of an outer skin, a vascular system to transport water and nutrients, supporting cells, and the ground tissue or flesh cells. The water content of vegetables is approximately 90%, and loss of the water causes the vegetables to wilt. Cell walls are composed primarily of cellulose, which can be softened but not dissolved by cooking, and pectic substances, which can be broken down by cooking. Neither cellulose nor much of the raw pectic substance can be digested, making vegetables a very important source of bulk and fiber in the diet. Another component of vegetables that cannot be softened or digested is lignin, found in the woody or stringy portions.

COLOR. The color of vegetables is due to the various pigments they contain. Four important pigments are the green pigment chlorophyll, the yellow to orange-red pigment carotinoid, the red pigment anthocyanin, and the white pigment anthoxanthin. Often a change in pH (acidity) or overheating during cooking causes vegetables to change color.

The pH number is a measurement of the acidity or alkalinity of a solution. Pure water has a neutral pH of 7, while tap water has a slightly higher pH, indicating it is more alkaline, or *basic*. Most vegetables have a slightly acidic pH of 5–6. Ingredients that lower the pH, that is, make the vegetable mixture acidic, are vinegar, tomato juice, lemon juice, wine, and cream of tartar. The addition of baking soda would make a high, or alkaline, pH.

FLAVOR. The distinctive flavors of some vegetables are due to their sulfur compounds. For example, sulfur adds to the flavor of cabbage, cauliflower, broccoli, mustard greens, and turnip greens.

Classification

One classification of vegetables is based on the part of the plant that is eaten. For example, the roots of carrots are the part usually eaten, so carrots are classified as a root vegetable. Vegetables can also be classified by their color, flavor, or moisture content.

Characteristics	Examples
Edible part of plant	
Root	Beet, carrot, radish, parsnip
Tuber	White potato, baking potato, and red potato
Bulb	Onion, garlic, leek, shallot
Stem	Celery, asparagus
Leaf	Spinach, lettuce, parsley, cabbage, Brussels sprout
Flower	Cauliflower, broccoli
Seed and pod	Legumes (lima bean, dried pea, kidney bean), sweet corn, green bean, sweet pea, wax bean
Fruit	Green and red peppers, summer squash, cucumber, tomato, eggplant, winter squash, okra
Color	
White	Onion, potato, cauliflower
Green	Lettuce, spinach, green pepper, broccoli
Yellow	Carrot, corn, rutabaga
Red	Beet, tomato, red cabbage, red onion
Flavor	
Strong	Broccoli, cabbage, onion, cauliflower, turnip, Brussels sprout
Mild	Asparagus, bean, beet, parsnip, carrot, corn, lettuce, mushroom, potato
Moisture content	
High	Lettuce, radish, tomato, onion, asparagus, cabbage, carrot, green and red peppers
Low	Okra, parsnip, potato, acorn squash
Dry	Dried bean, lentil, dried onion, dried parsley

SELECTION

Whatever the classification of vegetables, an acceptable product can be expected only when high-quality raw, frozen, canned, or dried products are used. Vegetables should be fresh when purchased or processed. Starting with the appropriate grade that indicates the quality of vegetable is the best way to ensure correct quantity, yield, and the best-quality end products. Therefore, the grade of vegetable that is needed by the foodservice operation for a particular menu item should be decided before purchasing. For example, vegetables used in soups, stews, or for pureed products may be a lower grade.

Fresh Vegetables

When buying fresh vegetables, check these characteristics:

- Ripeness: Vegetables will not ripen in storage. If they are not ripe when they are purchased, they will have poor texture and flavor.
- Color and texture: Vegetables should have a bright, characteristic color and a crisp texture. Color is the key to nutrition, especially for vitamin A content.
- Shape: Misshapen vegetables may be poor in texture and flavor.
- Size: Extra-large vegetables may be overripe, coarse, and have poor flavor. Extremely small ones tend to be immature and have poor flavor, too. Vegetables should feel heavy in relation to their size.

In addition to these points, this purchasing guide for fresh vegetables shows the characteristics of specific vegetables that indicate good quality.

Vegetable	What to Look For	What to Avoid
Asparagus	Compact, flexible tips; smooth, round spears; rich green color; crisp, tender stalk	Overgrown size; open and spread-out tips; no flexibility; excessive amount of sand
Bean, green	Bright color; young, tender, crisp	Thick, tough, fibrous; overripe
Beet	Rich, deep red color; firm, round root; fresh green top	Wilted; soft; elongated shape; tap root missing
Broccoli	Firm, compact cluster of small buds; dark green or sage green color; fresh stem	Spread cluster, open or enlarged bud; yellowish green color; slippery, water-soaked spots; soft
Brussels sprout	Fresh, bright green color; firm, tight-fitting outer leaves	Yellowish green color; signs of injury from worms
Cabbage	Hard, firm head	Overripe; worm-eaten wrappers
Carrot	Well-formed; smooth skin; firm; bright color	Large sunburn spots; flabby; wilted; overripe
Cauliflower	White color; compact, solid, clean bud; fresh green-colored jacket leaves	Spread or wilted bud; discolored spots; smudgy or speckled appearance
Celery	Solid, rigid texture; glossy light green color	Insect injury; brown or gray inside; soft; overdeveloped
Corn	Juicy; plump; bright yellow, small-size kernels; good green-colored husks	Worm injury; overripe; wilted or dried husks

Vegetable	What to Look For	What to Avoid
Cucumber	Good green color; well-shaped; firm; small lumps on the skin	Extra-large size; dull yellowish color; soft, wilted end
Eggplant	Firm, heavy, smooth; dark purple color with shiny peel	Overgrown size; soft; irregular dark brown or white spots on the surface
Greens	Young, tender texture; good green color	Coarse, fibrous stem; yellowish-green color; evidence of insects on the leaves; wilted leaves
Lettuce	Crisp, good bright color for variety; medium size	Overripe, decay; irregular shape or hard bump on top; small for weight
Okra	Tender pods that bend with pressure; under 4½"; fresh dull-green color	Tough or fibrous pods; pale or faded green color; soft or split skin
Onion	Firm, dry with small neck; covered with papery scales; free from sunburn or blemish	Wet or with soft neck; fresh sprout; thick center in the neck
Parsnip	Small or medium; firm; well-formed; smooth; free from blemish	Large; coarse; yellow; decay
Pepper, sweet green or red	Dark green or red color with glossy surface; firm wall	Thin wall; soft watery spots on the surface
Potato	Well-shaped; firm; no blemish or damage	Sprouted; shriveled; green color; bruised; soft
Scallion, shallot, and leek	Fresh, crisp, green tops; 2–3" of tops from the root end	Yellow color; wilted; bruised tops; discolored tops
Sweet potato	No blemishes; firm; no worm holes	Worm holes; cuts

Processed Vegetables

CANNED VEGETABLES. The advantages of using canned vegetables are moderate cost, year-round availability, consistent quality within brands, and good variety. In addition, the amount of energy needed in cooking is less than for fresh. Canned vegetables are produced according to the U.S. government standards and may have a grade on the label. Grading services are provided by the U.S. Department of Agriculture (USDA). The grade may be listed as a letter or descriptive word. Only when *U.S.* is before the grade, such as U.S. Grade A, has the product been inspected and graded by the USDA. If *U.S.* does not appear with the grade, the product has not been federally inspected. Standards for canned vegetables are as follows.

Grade	Quality of Vegetables
U.S. Grade A, or Fancy	Best-flavored, tenderest and juiciest; uniform in size, shape, color, and tenderness; represents choice of crop
U.S. Grade B, or Choice; Extra Standard	Flavor fine, tender and juicy, may be slightly more mature, firmer in texture, sometimes less uniform than fancy grade
U.S. Grade C, or Standard	Flavor less delicate; firmer in texture; often less uniform in size, shape, color; more mature

The grade purchased should depend on the intended use, with the lower grade being used in mixed-food combinations such as soups.

Avoid using canned vegetables with any signs of spoilage, such as cans that bulge at the sides or ends, liquid spurting out of the can when it is opened, mushy contents, or a sour smell. In all of these cases, *NEVER* taste the food. It may contain a toxin that causes botulism, which can be fatal even in a very small amount. Discard all canned products with any of these signs. Remember, a canned vegetable that is spoiled may look and smell all right but still be unsafe.

FROZEN VEGETABLES. The three grades of frozen vegetables are U.S. Grade A, or Fancy; U.S. Grade B, or Extra Standard; and U.S. Grade C, or Standard. The grade is not commonly listed on the label. Frozen vegetables provide freshness and vivid colors while saving labor time and limiting waste, compared with the preparation of fresh vegetables. They are closest to fresh in nutrients, color, and flavor. Frozen vegetables are available whole or in cut pieces.

Some frozen vegetables are packed in special sauces, such as cheese or butter sauce. Frozen vegetables for quantity production usually come in 2½-pound boxes or in plastic bag–lined cases of 20–30 pounds. Frozen vegetables must be frozen thoroughly. Packages with holes, tears, open seams, or other damage should be avoided because they can be contaminated. Vegetables that have been thawed and refrozen will have a lower quality and poorer texture.

DRIED VEGETABLES. Legumes (lentils, peas, beans, and soybeans) are usually available in a dried form. Some vegetables and herbs that are commonly used fresh are also used dried for seasoning, such as chives, basil, oregano, onion, and parsley. The usual way to produce mashed potatoes in quantity food production is from potato flakes or granules. Look for the following characteristics when buying dried vegetables.

- Uniform, bright color. Loss of color means staleness and longer cooking time.
- Uniform size. In legumes, mixed sizes will cook unevenly because smaller pieces cook faster than larger ones.

In addition, there should be no visible defects, for example, the presence of foreign materials such as sticks, stones, or dirt. The container should be tightly sealed.

STORAGE

Vegetables should be stored in optimal conditions to retain their nutrients, flavor, and texture. Here are suggested storage temperatures and conditions for different types of vegetables.

Vegetable	Temperature (F)	Storage Conditions
Frozen vegetable	0 to −20	In original container or in moisture- and vapor-proof wrapping (heavy-duty foil or heavy-duty plastic container with tight-fitting lid)
Leafy vegetable	45	For 7 days, unwashed
Potato, onion, root vegetable	70 (room temperature)	For 7–30 days; keep in ventilated bag or container
Fresh vegetable, such as corn, green bean, pea	Refrigerate	In plastic bag or container approved by National Sanitation Foundation
Dried vegetable	Dry and cool	Once opened, store remaining portion in a plastic container with a tight-fitting lid

PREPARATION

Some foodservice operations use packaged fresh vegetables ready for cooking when purchased. These vegetables have been washed, peeled, and/or cut prior to purchase. This saves the costs of equipment and labor in preparation. However, these fresh convenience vegetables are relatively high in cost, may have a less desirable flavor, and may have preservatives added. For these reasons, many foodservice operations prefer to maintain quality standards by purchasing fresh vegetables without prepreparation.

Fresh Vegetables

Most fresh vegetables spoil easily and should be refrigerated as soon as they are received. Some large-quantity operations use an ingredient room or a vegetable-preparation area. However, many nursing facilities are not large enough to make these areas cost effective.

Proper sanitation methods are necessary when handling vegetables. It is very important to realize that clean vegetables have a much longer shelf life than do dirty and partially spoiled ones.

Raw vegetables should be washed thoroughly before use. Soaking the entire vegetable in cool water during cleaning also helps dried or wilted vegetables absorb water. However, some vegetables, like mushrooms, become soggy if soaked too long. It might be necessary to use a special technique to loosen extra dirt. One technique is to wash the vegetables in room-temperature water. This method may cause slight wilting, but vegetables can be recrisped by rinsing with cold water. Another technique is to add a small amount of salt, about 1 tablespoon per gallon of water, to help loosen dirt and grime on leafy and flower vegetables. This also helps remove small insects from inner leaves. This method should not be used for preparing vegetables for low-sodium diets.

All leafy vegetables should be washed under cold running water and drained thoroughly at least 2 hours before use. Some leafy vegetables, such as spinach and leaf or endive lettuce, need a more thorough cleaning. The quickest and most efficient way to clean the veins of greens is to fill the sink with cold water and slosh the leaves up and down several times to remove the dirt. While draining the sink and filling it with fresh, cool water, the greens are drained on a rack. The sloshing action and draining are repeated until the water remains clean. After washing, the greens should be drained and stored in plastic bags or airtight containers and refrigerated to retain freshness and crispness. See Figure 3.1.

Soaking vegetables too long will cause nutrients to be lost. Overhandling of some vegetables, like lettuce, will increase the rate of spoilage due to bruising of the leaves, so minimum handling is recommended.

Vegetables usually require special handling. After washing, they should be carefully inspected to remove all blemishes for quality control. At this point, any necessary trimming, paring, and cutting should be done.

Trimming is done to remove woody stalks; heavy, tough leaves; and other parts that are not eaten. Correct trimming depends on the vegetable. For example, green beans should be trimmed at both ends to remove the tough cellulose area of the steam and blossom, cabbage wedges should be cut so the remaining core will hold the wedge together, and beets should not have their roots cut off before cooking or they will not keep their color. Bruised and spoiled sections of all vegetables should be trimmed immediately. It takes very little time for one spoiled vegetable in a case to increase the spoilage rate of the remaining ones.

To avoid bruising vegetables, it is important to use sharp stainless steel knives when cutting them. In addition, stainless steel will not discolor by interacting with the acids in the vegetables.

Paring is done to remove skins of some vegetables, like potatoes and carrots. Because many minerals and vitamins in vegetables are found in or near the outer skin, paring should be as close to the surface as possible. Use of a knee-action peeler will minimize the paring loss. When a

Figure 3.1. Procedure for washing lettuce.

commerical potato peeler is used, the amount of peeling must be watched and timed carefully. Excessive or deep peeling by the machine will increase the loss of minerals and vitamins and waste some of the vegetables, increasing food cost.

Variations in food yield depend on the kinds of vegetables and the tools and methods used in trimming or paring. The following table (from Matthews and Garrison, *Agriculture Handbook* No. 102, 1975) gives the approximate yield after the preparation of certain vegetables.

Vegetable	Yield (%)
Asparagus	93
Broccoli, trimmed	78
Cabbage, green	79
Carrot, topped	82
Cauliflower	55
Onion, mature, white	81
Potato	84

To determine the amount to purchase of fresh vegetables, divide the amount needed after cleaning and preparation, called the edible portion (EP), by the food-yield factor. The food-yield factor is equal to the yield divided by the original weight times 100. For example, if there are 8 pounds of peeled, trimmed carrots in a recipe, an amount must be purchased that will yield 8 pounds of peeled, trimmed carrots. To determine the as-purchased (AP) weight of carrots required, the 8 pounds are divided by 0.82 (82% yield ÷ 100). Therefore, to serve 8 pounds:

$$EP \div yield = AP$$
$$8 \div 0.82 = 9.756 \text{ lb} = 9 \text{ lb } 12 \text{ oz}$$

So 9 pounds 12 ounces of raw carrots are needed. When the amount of AP is known, but not the EP, the following formula is used:

$$AP \times yield \% = EP$$

COOKING FRESH VEGETABLES

Boiling. Boiling and steaming are the most commonly used methods in quantity vegetable cookery.

Boiling may be done in an open kettle or in a steam-jacketed kettle. Large or small amounts of vegetables can be cooked evenly in a steam-jacketed kettle with a wire insert basket. Boiling methods should be selected to retain the bright color and shape of the vegetables (see Table 3.1). After cooking, the vegetables should be drained quickly to avoid overcooking.

Steaming. Most quantity foodservice operations use commercial compartment steamers. Usually vegetables can cook more evenly by steaming rather than boiling. When using a steamer, vegetables are placed in single layers to cook rapidly and to prevent crushing. Vegetables should be steamed until tender-crisp. Small, high-pressure steamers are used for the smaller number of servings in modified diets in nursing facilities and hospitals.

Newer compartment steamers give the operator a choice between pressureless and vented or pressure and unvented steam cooking. Vegetables that can be cooked at low temperatures, uncovered, on a range can be cooked by pressureless, vented steam. These include peas, carrots, snow peas, Brussels sprouts, spinach, green beans, and cauliflower. A compartment steamer can keep nutrients better than any other cooking method because of shortened cooking time and the small amount of water used.

The compartment steamers use solid and perforated pans. The perforated pans are used for vegetables like snow peas that are steamed without water in the pan. Care should be taken to fill the pan evenly. Since the product should not be cooked until it is needed for service, it is best to divide the total vegetable production into smaller batches. This should help ensure that nutrients, texture, flavor, and color are retained. If the product is overcooked or held more than a few minutes after cooking, flavor, color, texture, and nutritional value decrease. See Table 3.1 for steaming times for selected fresh vegetables.

Baking. *Baking* (cooking by dry heat in the oven) is used for cooking potatoes, squash, tomatoes, and eggplant. Baking retains more vitamins and minerals than most other cooking methods. Except for green vegetables, vegetables are baked in a covered pan. Vegetable casseroles also are prepared by baking. Temperatures for baking vegetables usually range from 300 to 425 F.

Stir-frying. Stir-frying is an Oriental cooking method that can be done in a skillet, fry pan, or steam-jacketed kettle in quantity foodservice operations. Vegetables must be sliced into thin, uniform pieces and cooked until the outside is just done. In spite of their thin slices, the vegetables are still crisp and moist after cooking. The relatively short cooking time reduces the loss of water-soluble nutrients, such as vitamin C. Other advantages of stir-frying are the retention of the fresh flavors and vivid colors.

Microwave Cooking. Microwave cooking is an ideal method for cooking small amounts of vegetables. Because the microwave oven cooks quickly and little water is used, vegetables retain most of their color, flavor, nutrients, and texture. Microwave cooking is discussed thoroughly in Chapter 11.

Processed Vegetables

COOKING FROZEN VEGETABLES. Most frozen vegetables have been partially cooked or blanched, so the final cooking time is shorter than for fresh vegetables. If vegetables are frozen in a solid block, like spinach, they should be partially thawed in the refrigerator and/or broken into pieces before cooking to will help them cook more evenly.

Table 3.1. Techniques for boiling and steaming selected fresh vegetables

Vegetable	Use of Cover	Reason to Cover or Uncover	Amount of Water	Size of Piece	Boiling Time (min)[a]	Steaming Time (min)[a]
Asparagus, whole	Uncovered	Green color	Small[b]	Tips	5–10	7–15
				Stalks	10–20	12–30
Beans, green	Uncovered	Green color	Small	Whole	20–30	25–35
Beans, wax	Uncovered	Yellow color	Small	Whole	25–30	25–35
Beets	Covered	Mild flavor	Small	Whole	30–90	40–90
Broccoli	Uncovered	Green color, strong flavor	To partly cover[c]	Split stalks	10–20	15–20
Brussels sprouts	Uncovered	Strong flavor	To cover	Quartered or shredded	10–20	15–20
Cabbage, green	Uncovered	Green color, strong flavor	To cover	Quartered or shredded	10–15	15–20
					3–10	8–12
Cabbage, red	Uncovered	Strong flavor	To cover	Shredded	8–12	10–15
Carrots	Covered	Mild flavor	Small	Small, whole	15–20	20–30
Cauliflower	Uncovered	Strong flavor	Large	Whole or flower	15–25	25–30
					8–15	10–20
Corn on the cob	Covered	Mild flavor	To cover	Whole	5–10	10–15
Onions	Uncovered	Strong flavor	Large	Whole, small, or whole, large	15–30	25–35
					20–40	35–40
Parsnips	Uncovered	Strong flavor	Large	Whole or quartered	20–40	30–45
					8–15	30–40
Peas	Uncovered	Green color	Small	Whole	15–20	5–6
Potatoes	Covered	Mild flavor	Small	Whole	25–40	20–30
Spinach	Covered to wilt then uncovered	Green color	Clings to leaves	Leaves	3–10	5–12
Sweet potatoes	Covered	Mild flavor	Small	Whole	30–40	20–30

Source: Adapted from McWilliams (1985).
[a]Variations in cooking time depend on the state and maturity of vegetables, as well as personal preference.
[b]Just enough water is used to bubble to the top of the vegetable when the water is boiling gently.
[c]Water to within ¼" of flowers when broccoli is standing upright.

Steaming is the recommended cooking method for frozen vegetables in quantity foodservice preparation. A full-size, 12×20×2½" pan will hold a 2- to 4-pound package of vegetables. Standardized recipes should give directions on how to cook frozen vegetables.

COOKING CANNED VEGETABLES. Canned vegetables are fully cooked when canned and need only heating for serving, 140–160 F. Overheating results in further loss of nutrients, a soft texture, an unattractive appearance, and a poorly flavored product. The dull-green color of canned green vegetables is due to the long cooking at high temperatures necessary for safety during the canning process.

The liquid from canned vegetables contains water-soluble nutrients and should be used in cooking. Since only enough liquid to cover the bottom of the pan is needed when heating canned vegetables, any remaining liquid may be used in preparing soups, stocks, and sauces. Because this liquid can be high in salt, additional salt may not be necessary when it is used in other products. Vegetables should be drained and seasonings added, if needed, before portioning.

COOKING DRIED VEGETABLES. Dried legumes, such as navy, lima, pinto, and kidney beans, must be soaked before cooking to shorten the cooking time and increase the yield and digestibility. The length of the soaking period depends on the temperature of the water. The soaking water can be used in cooking dried vegetables.

Dehydrated onions are one of the most common types of dried vegetables used in quantity foodservice operations. Usually 1 pound of dehydrated onions can be substituted for 8 pounds of fresh onions. To rehydrate the onions, cover them with cool or lukewarm water 1½ times the volume of the dried onions and let them stand 20–30 minutes until the water is absorbed. For example, if a recipe requires 1 pound of fresh onions, substitute 2 ounces of dry onions and soak in ½ cup of room-temperature water.

CONTROLLING QUALITY CHANGES DURING COOKING

Preparation of high-quality vegetables can be challenging in quantity foodservice. It is important to choose a preparation method that can maximize the desirable characteristics of the vegetables. An appealing vegetable may be ruined by incorrect cooking methods. During the cooking process, the color, texture, and flavor may be altered and some of the nutrients lost. The quality of cooked vegetables is determined by the amount of change in texture, color, flavor, and nutrients. Overhandling or stirring during cooking can result in misshapen or broken vegetables. Vegetables should be undercooked rather than overcooked, to ensure maximum retention of nutrients as well as desirable texture. Successful mastery of vegetable cookery is a must if meals high in nutritive value with a variety of textures, flavors, and colors are to be offered.

Effects of Cooking on Texture

The cooking process causes a breakdown of protein, thickening of starch, and softening of cellulose, pectic substances, and lignin in the vegetable structure. These physical changes result in a change in texture, with the final effect depending on the cooking time. The cooking time is determined by the type and amount of vegetables, amount of cellulose, presence of acids, size of pieces, desired degree of doneness, and cooking method used, that is, pressure or pressureless. The amount of fiber varies with the type of vegetable and maturity. In some vegetables, such as asparagus and broccoli, the amount of fiber even varies between the stem and the bud. Buds of broccoli or asparagus contain less fiber than do stems, so they need less cooking time.

The acidity or alkalinity of the cooking water influences the texture change of vegetables. Vegetables quickly become mushy in alkaline water. A small amount of baking soda causes the water to become alkaline and destroys the water-soluble vitamins. Therefore, baking soda is not recommended in vegetable cooking.

On the other hand, acid cooking water increases the firmness of cooked vegetables. An acid ingredient such as tomato juice in vegetable soup slows the breakdown of firm structures during cooking. Small pieces of vegetables in vegetable soup may retain their shape even after an hour of cooking.

The desired degree of doneness also varies with different types of vegetables. Winter squash and potatoes are considered done when they are quite soft, but most vegetables should be cooked until they are tender-crisp. Vegetables at this stage of tenderness are not only more palatable but they retain maximum color, flavor, and nutrients. One exception is the ideal tenderness of green beans. Many people prefer green and wax beans to be tender but firm.

Cultural differences influence preference for vegetable tenderness. Therefore, clients' preferences as well as nutrient quality should be considered in production.

Effects of Cooking on Color

Color is a major appeal of vegetables, so the cooking time and amount of liquids should be selected to retain color. Optimal cooking methods usually differ with the color pigment that dominates in the vegetable. Only when the color is stable during cooking should vegetables be cooked in a covered pan. Table 3.2 is a summary of changes in color of vegetables under varying conditions.

Table 3.2. Changes in color of vegetables under varying conditions

Pigments	Color	Effect of Acid	Effect of Alkali	Water Solubility	Effect of Overheating
Chlorophyll	Green	Olive green[a]	Bright green	Slightly water soluble	Olive green[a]
Carotinoid	Orange	Orange	Orange		
Anthocyanin	Red	Red	Blue,[a] green	Water soluble	
Anthoxanthin	White	Colorless, white	Yellowish[a]	Water soluble	Gray[a] or yellow

[a]Undesirable color for vegetable quality.

GREEN. The pigment in green vegetables is chlorophyll. Proper cooking should brighten the green color, but overcooking causes green vegetables to turn a dull olive green. In addition, covering green vegetables during cooking turns them dull olive green because the natural acid of the vegetable cannot escape.

Fresh and frozen green vegetables should be cooked in a steamer or in a small amount of boiling water to retain their green color. Adding an alkaline like baking soda may produce a bright green color, but as mentioned earlier, it should never be used because it will destroy most of the water-soluble vitamins, increase the sodium content, and produce a mushy texture.

Steaming is the recommended method of cooking green vegetables in quantity foodservice operations. Timing must be controlled carefully to achieve optimal color and flavor. Overcooking or holding too long before serving will cause a change from bright green to a dull-green, unappealing color.

Many quantity foodservice operations use compartment steamers in vegetable cookery. Vented, pressureless steamers are preferred for all green vegetables because they allow the steaming of vegetables to be done uncovered.

YELLOW. The pigment in yellow vegetables is carotinoid, a source of vitamin A. This pigment is very stable in water and heat but soluble in fat. Because yellow or orange carotinoid-containing vegetables are not affected by the acidity or alkalinity of the cooking water, these vegetables can be cooked with or without a lid. In quantity food production, they are cooked in a pressure or pressureless compartment steamer or a steam-jacketed kettle.

WHITE. The pigment in white vegetables is anthoxanthin. It stays white in acid but turns yellow in alkaline solutions. The addition of a little bit of acid near the very end of the cooking time can keep the vegetable white without lengthening the cooking time appreciably, for example, adding 1 teaspoon of lemon juice to 1 pound of cauliflower. White vegetables should be cooked covered to keep the volatile acids in the pan and to keep the color unchanged. Steaming for a short period helps maintain the color and flavor of white vegetables. Overcooking or holding too long in the steamer or on the tray line may turn them dull yellow or gray.

RED. The pigment in red vegetables is anthocyanin. It is stable in acid but changes to blue or purple in an alkaline solution. In making sweet-sour red cabbage, the addition of vinegar enhances the red color. Only a small amount of vinegar should be added right away and the rest after the vegetable is tender because adding the full amount of vinegar would toughen the cabbage and extend the cooking time. Although the color would be enhanced by covering the cabbage, optimal flavor is obtained by cooking uncovered.

Because of its alkalinity, hard water can change red vegetables to an unattractive blue or bluish green. Anthocyanin pigments are placed at the end of the roots and are water-soluble, so fresh beets should be cooked unpeeled, untrimmed, and in an uncovered pan. If roots are removed so less than 1" of stem remains, the red color will be lost into the cooking water. Therefore, most quantity foodservice operations use canned beets.

Effects of Cooking on Flavor

Although color plays an important role in acceptance of vegetables, flavor and aroma are also key aspects in making vegetables appealing. The aroma of cooked or fresh vegetables adds to the flavor of food. Most flavor-producing substances are water-soluble, so they can easily be lost or dissolved while cooking in water. A longer cooking period will make the flavor of some vegetables milder because some of the flavoring compounds are lost into the water or as steam.

Strong-flavored vegetables such as onions and cabbages will have a good color and mild flavor if cooked correctly. The onion family becomes milder after cooking and the desirable aroma can be achieved by cooking onions for a moderate length of time. On the other hand, the cabbage family develops intense flavors with longer cooking time, the longer the time, the more undesirable the flavor. A short cooking time without a lid achieves the best flavor in cabbage. It should not be held in the steamer or on the tray line longer than 20 minutes.

Proper cooking methods will bring out or improve the flavor of mild-flavored vegetables. Even slight overcooking of mild-flavored vegetables does some damage. A minimum amount of water along with a short cooking time maintains the flavor and taste of mild-flavored vegetables.

Other factors believed to influence the effect of cooking on flavors are the hardness of the water and the presence of salt in the cooking liquid. The use of hard water in cooking increases the cooking time by hardening the cell structure of mild-flavored vegetables. Only the minimum amount of salt that is needed to develop and maintain good flavor should be added.

Here are choices to consider when selecting the best cooking method:

For Steam Cooking	For Range-Top Cooking
Pressure or pressureless	Covered or uncovered
Vented or unvented	Amount of liquid
Perforated or solid pan	Length of cooking time
Addition of liquid	Batch size
Length of cooking time	
Batch size	

Effects of Cooking on Nutritional Content

Nutrients in vegetables can be lost easily during cooking. Many vitamins and minerals are water-soluble, including the B vitamins and vitamin C, which can be destroyed in an alkaline solution. Other vitamins can be destroyed by either high heat or overcooking. *The methods used to retain good flavor, texture, and appearance also tend to be the methods that conserve nutrients.*

Often, a compromise in cooking methods is necessary. For example, increased amounts of water in the cooking process might improve the palatability of strong vegetables, but it increases the loss of sugars, other mild-flavored compounds, and water-soluble vitamins. Therefore, to conserve the nutrients, vegetables should be cooked in their skins or cooked only minimally to the desired degree of doneness. Excessive heating should be avoided to prevent additional losses of vitamins B and C. A balance between the amount of water used and the temperature selected must be considered to prevent the loss of these important water-soluble vitamins. Commercial antioxidants prevent undesirable color changes, but for safe use, package directions must be followed very carefully. *Sulfite antioxidants are considered unsafe for use in food production.*

GENERAL RULES OF VEGETABLE COOKING

To ensure the best vegetable characteristics, these general rules should be followed:

- Handle vegetables carefully to avoid bruising.
- Cut vegetables to uniform sizes for even cooking.
- Cook strong-flavored and green vegetables uncovered.
- Do not use baking soda when cooking vegetables.
- Cook vegetables as close to serving time as possible.
- Cook vegetables in as small amount of water as possible.
- Keep cooking time as short as possible, only until the vegetables are tender-crisp.
- Hold cooked vegetables a minimum time before serving, never over 15 minutes.

POTATOES

In large-quantity food preparations, potatoes are cooked in a variety of ways. They have excellent

VEGETABLES

keeping qualities if they are stored at cool room temperature. U.S. potatoes may be classified into four groups according to appearance: round white, russet, round red, and long white. Storage ability, solid content, texture, and other characteristics influence use and preference.

Potatoes are divided into two types based on texture: (1) waxy all-purpose and (2) mealy. Round white and round red groups are the waxy type. Russet and long white are the nonwaxy type. Waxy potatoes are relatively high in sugar. These potatoes will hold their shape after cooking and are optimal for stews, soups, pan-fried potatoes, and potato salad. Mealy potatoes are high in starch and low in sugar. They are best for French fries, mashed or baked potatoes, and potato-thickened soup.

Preparation

Potatoes should be selected according to a purchasing guide (see the purchasing guide presented earlier). They must be cleaned thoroughly with water and a brush to remove soil and microorganisms. Discoloration of potatoes caused by an enzymatic reaction may develop while they are handled or pared. To prevent them from turning brown after peeling and trimming, potatoes can be covered with clean, cold water. Soaking in water, however, will reduce the vitamin C content. Commercial antioxidant products, except the sulfite type, can be also used to prevent browning. This surface discoloration usually disappears when the potato is cooked.

Storage

Storage temperatures can change the sugar/starch ratio over a period of time. If the storage temperature is lower than 45 F, the amount of sugar increases and starch decreases. This decreases the optimal quality of cooked mealy potatoes. The best storage place for potatoes is dry and well ventilated at cool room temperature, about 60 F. To prevent potatoes from turning green, they should be stored in a dark place. Under this condition, they could keep their quality for 1–2 weeks. If potatoes are stored in the refrigerator, the humidity may cause mold and spoilage.

The green parts of potatoes contain a bitter-tasting chemical that could be poisonous in sufficient quantity. It develops when potatoes are exposed to light during storage or handling. Areas of the potato that are sprouting also contain this chemical. If green parts and sprouts develop during storage, they must be cut off before cooking.

Cooking

BAKED POTATOES. The baked potato is a genuine American food. It should be fluffy, white, mealy, and almost dry in texture. In large-quantity foodservice operations, baked potatoes are not usually wrapped in foil. Foil tends to slow down the cooking process and produces a steamed product. It also adds labor and supply costs.

The optimal baking temperature for potatoes is around 400 F. The holding period should be limited to keep them at their best quality. Long holding periods for baked potatoes will result in a moist and soggy texture, yellowish-gray color, and sour flavor. Some operations use the microwave oven to solve holding problems. Potatoes are oven-baked until they are three-fourths done and then refrigerated. The final cooking is completed in the microwave oven. However, only 2 to 4 potatoes can be heated at a time. See Chapter 11 for complete information.

MASHED POTATOES. Mealy types of potatoes are used for mashed potatoes. Some facilities use dried mashed potatoes to save time and labor costs. To prepare dried mashed potatoes, the instructions on the recipe should be followed. Adding seasonings like salt and white pepper can improve the flavor. The potatoes should be moist, smooth, and thick enough to hold their shape and have a creamy white color.

Dehydrated potatoes may be purchased in granules or in flakes. They are also available with dried milk solids and with seasonings.

FRENCH FRIES. French fries can be prepared in different cuts. Most quantity foodservice operations use frozen French fries. They are partially cooked and are prepared by deep-fat frying while still frozen. A good French fry is golden brown, cooked through, nongreasy, crisp on the outside, mealy-tender on the inside, and hot when served. To prevent French fries from becoming moist and soggy, they must be held only a very short time before serving.

DUCHESSE POTATOES. Duchesse potatoes are a type of mashed potatoes piped through a pastry bag. The mixture is prepared from hot mashed potatoes with egg yolks added. Only pasteurized egg yolks should be used because it is difficult to cook the yolks thoroughly once they are added to the potatoes and egg yolks must be cooked above 140 F. The potato mixture can be piped into different shapes and stored in the cooler for a few days or freezer for weeks before finishing the cooking. A duchesse mixture is also available in instant form.

Deep-fat-fried duchesse potatoes are golden brown and crisp on the outside with a soft yellow-white center. A good breaded duchesse potato has a light brown, glossy crust and soft texture inside. Mashed potatoes, whether plain or duchesse, should be smooth, free from lumps, fluffy, and mealy. Plain mashed potatoes should have a creamy white color with no hint of gray, while duchesse potatoes should be light yellow.

SUMMARY

Vegetables add color, shape, texture, and nutrients to meals. They are available in fresh, canned, frozen, and dried forms. Fresh vegetables are among the best sources of vitamins and minerals. Vegetables may be classified by the parts of the plant eaten, the color, the flavor, or the moisture content.

High-quality vegetables must be purchased to have high-quality food products. Guidelines should be followed when buying, preparing, and storing vegetables.

Cooking may improve the color and texture and change the flavor and nutritional value of vegetables. It is important to choose preparation and cooking methods that can maximize the desirable characteristics of vegetables.

LEARNING ACTIVITIES

Activity 1: Preparation of Fresh Vegetable Medley

1. In a quantity foodservice laboratory or in an institutional setting, prepare the Fresh Vegetable Medley recipe for 32 portions.

Fresh Vegetable Medley

.... Yield: 96 portions
           3 pans 12×20×2½"
.... Portion size: #8 dipper *or* 3 oz

Amount	Ingredients	Amount	Procedure
....	Cabbage, shredded, medium coarse	6 lb 8 oz	1. Prepare vegetables by washing, trimming, and slicing.
....	Carrots, sliced ⅛"	1 lb 12 oz	2. Cook vegetables in steamer.
....	Celery, sliced		*Note:* When served, the vegetables
....	crosswise ⅓–½"	2 lb 4 oz	should be tender but retain a suggestion
....	Onion rings, sliced thin	2 lb 4 oz	of crispness. DO NOT OVERCOOK.

	Green pepper rings, sliced thin	1 lb 4 oz	a. Weight cabbage and carrots into 3 12×20×2½" pans; steam. *Approx time:* 4 min (Vegetables will not be completely cooked.)
			  b. Add celery to combined cabbage-carrot mixture; steam. *Approx time:* 3 min
			  c. Add onion and green pepper rings to cabbage-carrot-celery mixture; steam. *Approx time:* 5 min
			  d. Drain, if necessary.

....	Tomatoes, fresh	4 lb 4 oz	3. Blanch tomatoes; peel and cut into quarters or, if large, into 6–8 wedges each.
....	Margarine or butter	15 oz	4. Melt margarine or butter in large skillet or saucepan.
....	Salt	2 oz	5. Add tomatoes, salt, and pepper to margarine; cook over low heat only until tomatoes are heated.
....	Pepper, black	1½ tsp	
	Total weight	17 lb 12 oz to 18 lb 12 oz	6. Add tomatoes to the steamed vegetables. Mix lightly to distribute. 7. Serve with a slotted or perforated spoon.

Purchasing Guide

Food as Purchased	For 96 Portions	For ____ Portions
Cabbage, approx	8 lb 8 oz	
Carrots, topped, approx	2 lb 8 oz	
Celery, approx	2 lb 12 oz	
Onions, mature, approx	2 lb 12 oz	
Green peppers, approx	1 lb 12 oz	

Source: Adapted from *Standardized Quantity Recipe File* (1971).

A. Get the AP weight of each vegetable from the purchasing guide.

B. Prepare the vegetables by washing and trimming. Get the food yield and compare it with the table in the chapter. Drain the vegetables well and cook them. (If you have more of an ingredient than the amount needed, store the rest in the way that prevents wilting.)

2. Use the as-purchased and food-yield charts to compare the food yield. Remember: The actual food yield = (EP weight ÷ AP weight) × 100%.

Activity 1, Step 2

Foods as Purchased	For 96 Portions	For 32 Portions
Cabbage	8 lb 8 oz (approx)	
Carrots, topped	2 lb 8 oz	
Celery	2 lb 12 oz (approx)	
Onions, medium	2 lb 12 oz	
Green pepper	1 lb 12 oz	

Activity 1, Step 2

Vegetables	Recommended Food Yield	Actual Food Yield
Cabbage	80%	
Carrots, topped	82%	
Celery	90%	
Onions	91%	
Green pepper	85%	

VEGETABLES 55

3. Complete the product evaluation chart.

Activity 1, Step 3

PRODUCT EVALUATION CHART						
PRODUCT:						
Check the box that describes your rating of the product.						
CHARACTERISTICS	Excellent	Very Good	Good	Fair	Poor	Unacceptable
Appearance						
Texture						
Consistency of dressing						
Flavor						
Overall acceptability						

What techniques should be improved?

1. Measuring:

2. Temperature of vegetables:

3. Preparation:

4. Portioning:

Comments:

REVIEW QUESTIONS

True or False

1. Frozen vegetables require more cooking time than other forms of vegetables.

    A. True
    B. False

2. If vegetables are simmered or blanched for a long time, the texture will be very soft.

    A. True
    B. False

3. Most vegetables should be held in the steamer no longer than 15–20 minutes before serving.

    A. True
    B. False

4. The highest grade for canned vegetables is U.S. Fancy.

    A. True
    B. False

5. Cabbage should be cooked over high heat to get rid of its strong flavor.

    A. True
    B. False

Multiple Choice

6. Fresh potatoes should be stored in

    A. The refrigerator
    B. A cool, dry, and dark place
    C. A cool and moist place

7. Mealy potatoes are best for

    A. French fries and baked and mashed potatoes
    B. Stews and soups
    C. Potato salad and casseroles

8. Strong-flavored vegetables are

   A. Broccoli, onions, cabbage *(circled)*
   B. Lettuce, potatoes, zucchini
   C. Spinach, carrots, cucumbers

9. The pigment that gives color to yellow and orange vegetables and a source of vitamin A is

   A. Chlorophyll
   B. Anthocyanin
   C. Carotinoid

10. Which of the following is *not* true when cooking green vegetables?

    A. They should be cooked uncovered.
    B. They need baking soda to make their texture soft and their color bright.
    C. They should be cooked until tender-crisp. *(circled)*

11. Canned vegetables are

    A. Partially cooked
    B. Completely cooked
    C. Never cooked

12. All of the following statements relating to stir-frying are true, except

    A. Vegetables should be sliced into thin uniform pieces.
    B. Most quantity foodservice operations do more stir-frying than deep-fat frying.
    C. The color and flavor of vegetables can be retained.

Matching

13. Match the categories in column A with the vegetables in column B by writing the correct letter from column B by each number in column A.

Column A	Column B
__C__ 1. Root family	A. Tomato, cucumber, squash
__F__ 2. Tuber family	B. Cauliflower, broccoli
__D__ 3. Seed family	C. Beet, carrot, radish
__I__ 4. Bulb family	D. Green bean, pea
__B__ 5. Flower family	E. Spinach, lettuce, parsley
__G__ 6. Stem family	F. Potato
__E__ 7. Leaf family	G. Celery, asparagus
__H__ 8. Dry vegetables	H. Lentil, dried onion, dried parsley
	I. Onion, garlic, shallot

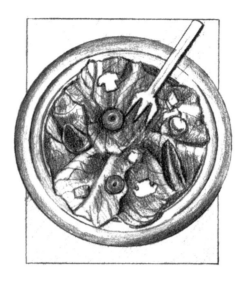

4. SALADS

Salads add variety to any meal through the flavors, colors, and textures of the individual ingredients. As a separate course or accompaniment to an entree, salads can contain ingredients from all of the food groups. Salads are also good sources of vitamins A and C and minerals.

UNDERSTANDING SALADS

Role of Salad in the Menu

Americans usually serve the salad at the beginning of the meal, as an accompaniment to the entree, or as the entree. Salads will vary in size and content according to their role in the meal. They are classified according to their course in the menu. A small, light salad served at the beginning of a meal is an appetizer. An entree salad contains a hearty, protein food and is a popular choice for lunch. An accompaniment salad is a light and refreshing side dish served with or before the entree. Usually this is larger than an appetizer salad but is not as large or hearty as an entree salad.

Standards for Salad

There are standards for planning all types of salads. These are guides for the appearance, color, flavor, texture, and temperature and how well the salad goes with the rest of the meal. A salad should be colorful and look fresh. The ingredients should be arranged simply on the plate or in a bowl. Each ingredient should be identifiable. The texture and flavor should enhance the rest of the meal. The flavor, whether zesty or merely pleasant, may come from either the ingredients or from the salad dressing. A salad should have some ingredients that are crunchy or crispy. A well-combined variety of textures will create an appetizing salad.

Salad preparation and style of service can require much time. As a result, foodservice managers should plan salad menus according to the workload requirements of the rest of the menu.

Parts of a Salad

There are four parts to a salad: the base, or liner; the body; the salad dressing; and the garnish.

BASE (LINER). A leaf of lettuce, endive, or cabbage on which the other ingredients are placed provides a contrasting color or texture and helps hold the salad in the center of the plate or bowl. Leaf, romaine, head, or Boston lettuce and red or green cabbage are commonly used as a base for a salad. (See the recipe for salad liners.)

Salad Liners			
Yield: 100 servings Portion size: as indicated			
Use	Ingredients	Amount	Procedure
Salad Plate	Lettuce, leaf	2½ lb (approx 4 head)	1. Wash greens gently but thoroughly using cool water. Several changes of water may be necessary in order to thoroughly clean the greens.
Nappy	Lettuce, leaf	1½ lb (approx 2–3 head)	
Luncheon Plate	Lettuce, leaf	5 lb (approx 8 head)	2. Trim carefully, removing bruised, damaged parts.
Salad Plate	Endive	3 lb (approx 3 bunches)	3. Drain greens well on a rack.
Garnish (small)	Endive	1 lb (approx 1 bunch)	4. Refrigerate to crisp and chill. Cover to prevent drying, but do not seal tightly.
Garnish (large)	Endive	2 lb (approx 2 bunches)	

Source: Laboratory Directions for Hotel, Restaurant, and Institution Management, Iowa State University, 1992.

BODY. The body is the main part of the salad. The body can be made from vegetables, fruits, poultry, fish, meat, or cheese.

SALAD DRESSING. The dressing usually is a combination of oil and vinegar or mayonnaise with a variety of seasonings. It provides flavor that goes well with the body of salad. It can be spooned over the salad, served separately, or mixed with the salad body before serving. Common examples are French dressing, Italian dressing, mayonnaise, or low-calorie dressings.

GARNISH. A garnish is sometimes used to providing color and texture contrast. It should be simple and not overpower the salad. Examples are sliced tomatoes, parsley, carrot curls, olives, or croutons.

TYPES OF SALADS

There are four basic types of salads: fruit, vegetable, protein, and gelatin.

Fruit Salad

Fruit salads may be served as an entree at a luncheon, as an accompaniment to the main course, or as a dessert. Fruits for salads may be fresh, canned, frozen, or dried. Usually frozen berries are used only in gelatin salads because they become mushy when defrosted.

FRESH FRUITS. Ripe fresh fruits can make colorful, attractive, and popular salads. Large fruits may be served in slices, cubes, or attractive pieces. For example, melons can be cut in wedges, cubes, or balls. The cut shape of the fruit should improve its appearance and natural shape. Fresh fruit that is cut into very small or odd-shaped pieces may lose its appeal. Pits should be removed from fruit before it is used for salad, especially in nursing facilities.

CANNED FRUITS. Canned fruits are frequently used in gelatin salads because they require no special preparation other than draining and chilling. Peaches, pineapples, pears, and mandarin oranges are popular. The best-quality canned fruits should be used (see Chapter 2).

DRIED FRUITS. Dried fruits also are used in some salads. Stewed dried fruit can be whipped with gelatin and cream to form a sweet fruit whip. Dried prunes, peaches, pineapples, and apricots also can be used as garnishes. See Chapter 2 for a discussion of dried fruits.

Vegetable Salad

Any vegetables—raw, cooked, frozen, or canned—can be used in making salads. Raw vegetables provide the most vivid color, texture, and flavor. The appearance of salads can be enhanced by the many styles of cutting vegetables: julienne, sliced, diced, and grated. The wide range of color, such as yellow corn, creamy white cauliflower, deep red tomato, and a variety of greens creates an appealing salad with visual interest.

Vegetable salads are good sources of nutrients and fiber and are quite low in calories. Mixed green salads provide fiber and vitamins to the diet.

How ingredients are combined is important in preparing a vegetable salad. For example, a cooked green bean salad needs the crunchy texture from diced celery, while white coleslaw will become a more colorful dish with grated carrots, diced green peppers or pineapple tidbits.

Molded Gelatin Salad

Molded gelatin salads are another popular type. They are fairly easy to prepare and add a variety of color and texture to the menu. Although the gelatin contains animal protein, it is in such a small amount that it cannot be considered as a source of protein. Gelatin salads are usually evaluated nutritionally for the content of ingredients molded into the salad. All types of gelatin can be used for salads: unflavored, sweetened flavored, and sugar-free flavored.

Protein Salad

When a protein, such as meat, fish, egg, cheese, beans, or poultry, is used in the body of salad, it can be an entree. These protein foods are used in combination with vegetables or fruits and salad dressing and are usually served on a leafy green liner. If protein salads are served as an entree, they need to include adequate protein and enough other ingredients to be the main course.

Ham is one of the most popular meats for salads because of its texture and smoky flavor. Chicken and turkey are also popular due to their delicate flavor and color. Seafood salads containing shrimp, salmon, and crab are more expensive sources of protein, whereas tuna is less expensive.

Other foods used in protein salads are cheese, eggs, and beans. Use of beans as a source of protein is discussed in Chapter 7. The yellow color of hard-cooked egg yolk is very attractive in salads. Chopped eggs add color and flavor to potato salad, sliced eggs can brighten tuna salad, and deviled eggs can be the main body of a salad.

When used in a protein salad, cheese can be combined with either vegetables or fruits. As a salad ingredient, cheese serves as a main source of protein. It can be diced, cubed, julienned, or shredded. Cottage cheese adds delicate flavor and combines well with either vegetables or fruits. Blue cheese may be an ingredient in salads or blended into dressing to give a sharp, distinctive flavor. Sometimes parmesan, a hard yellow cheese with sharp flavor, is sprinkled over tossed greens.

PREPARATION

The main rule for salad preparation is to keep the salad ingredients cold. All ingredients, including salad dressings, should be chilled before the salad is made.

Because many salad vegetables are not cooked, sanitary conditions during preparation and service are very important. Employees should use disposable plastic gloves or utensils when handling any vegetable after it has been cleaned and drained. Vegetable cleaning methods are explained thoroughly in Chapter 3.

Preparing Greens

Head and leaf lettuce should be washed completely and placed on a rack for draining. Excess water can be removed easily with a vegetable spinner. To use a spinner, the lettuce is put into the container, which is then spun with a handle or cord pull. The spinning motion removes droplets of water by centrifugal force, similar to the spin cycle of a washing machine. Overspinning removes too much water and will dehydrate the lettuce and cause it to wilt. The vegetable spinners come in two sizes. The smaller, manual type will hold one head, while the larger electric size will hold one or more cases of lettuce.

Leafy greens should not be stored in containers that are too small because some air is needed during storage and the leaves should not be crushed. A tightly sealed container or plastic bag creates a humid environment that allows the greens to *rehydrate,* that is, to reabsorb part of their normal water lost in shipping and storage. If the lettuce takes up the water it has lost, it becomes crisper and firmer. Complete exposure to the open air during storage will cause the vegetables to *dehydrate,* or to dry out and wilt.

For the best appearance and longest shelf life, salad greens should be torn by hand. However, because this may not be practical in a quantity foodservice operation, many use sharp stainless steel knives or a high-speed chopper to cut salad greens. It is important to remember that the more surface area cut and the greater the exposure to air, the faster the product will wilt and turn brown.

Preparing Molded Gelatin Salad

Molded gelatin salads can be prepared a day in advance. They can be molded in many sizes and shapes. Most quantity recipes specify that boiling water is added to flavored gelatin in a steam-jacketed kettle. If unflavored gelatin is used, the gelatin should first be rehydrated by soaking it in cold water (¼ cup water per 1 tablespoon gelatin) before heating in boiling water. Rehydration is necessary because the large gelatin particles absorb water more slowly than do the fine particles in flavored gelatin. After the gelatin has been rehydrated, the boiling water is added and the mixture stirred gently until dissolved. Then the remainder of the cold or even frozen liquid can be added.

To speed up the cooling and setting of the gelatin structure, ice can be used as a part of the liquid. Adding cold water or ice usually is done by slow stirring to ensure the gelatin is free from any thickened lumps. This rapid chilling weakens the final gel strength, so gelatin cooled by this method will melt more quickly at room temperature.

The ability of gelatin molecules to establish a network typical of a gelatin structure depends on its ability to be hydrogen-bonded together. This chemical change, hydrogen-bonding, is affected by the level of acid and sugar in the product. Addition of a large amount of vegetables, fruits, or other foods also weakens the gel. This happens because there is not enough gelatin to suspend the food and because any produce will add excess moisture to the mixture. Gelatin with mildly acidic ingredients will gel more quickly than when highly acidic, nonacidic or sugar, or very alkaline additions are made. With such additions, the gel will be softer because acidity affects the hydrogen-bonding, while additional sugar dilutes the solution.

Extra amounts of fruit juice will dilute the gelatin. Therefore, fruit juices must be considered as part of the total liquid called for in the recipe.

Raw, frozen, or fresh pineapple will prevent gelling as the result of *bromelin,* an enzyme. Papaya and kiwi have the same effect. For this reason, these fresh products should not be used in making molded salads.

If fruits and vegetables are to be added to the gelatin, they should be added once the gelatin is the thickness of unbeaten egg whites. This helps to distribute the fruits or vegetables evenly and also prevents the ingredients from sinking or from floating to the top of the pan. For a specific placement of fruit, the fruit is set in position in a thin layer of gelatin. After chilling, another layer is added to immerse the fruit in the gelatin. Fruits such as bananas require this additional layer of gelatin to keep them from turning brown.

An important factor to remember with gelatin preparation is that the gelatin must be completely dissolved in the solution. If the granules are not dissolved completely, a rubbery layer of gelatin will form at the bottom of the pan. Undissolved gelatin granules will give the final product the consistency of tapioca.

If an error is made in dissolving the gelatin granules or in the addition of fruits, vegetables, or liquids, the gelatin can be slowly warmed and returned to a liquid state. After correcting the problem, the gelatin can be refrigerated and returned to the gelled state.

Gelatin salad should be carefully unmolded before serving. A food-release product can be very lightly sprayed into the individual molds to help unmold the salad. Most gelatins can be removed from the mold by dipping the bottom half of the mold into warm water. The mold should not remain in the warm water longer than it takes to loosen the gelatin. Gentle shaking of the mold and turning it upside down will help free the gelatin. A plastic-gloved hand is used to catch the gelatin as it comes out of the mold. The leaf lettuce base that helps prevent the gelatin from sliding off the plate also hides any melted product. After the gelatin has been removed from the mold, it is returned to refrigeration, allowing the salad to become firm again.

Preparing Protein Salad

Because raw protein foods can be the source of cross-contamination, proper care is needed in preparation. Any cutting board and slicer should be sanitized before *and* after use. The ingredients should be chilled before and after preparation, and should be kept cold until serving time.

Tools Used in Preparing Salad

In making salads, proper tools can save time and energy and improve quality. A sharp stainless steel knife or stainless French knife and cutting board are used to chop or slice salad ingredients, such as fruits, vegetables, and meats. An egg slicer is another useful tool. Measuring equipment, including dippers and ladles, is necessary for portion control of salad dressings. Salad dressings are also available in portion-control packets.

PLATING SALADS

Salads are appealing if some attention is given to detail when they are served. A cold salad should be chilled before serving, all the ingredients should be arranged attractively on a chilled plate, and the proper amount of dressing should be added right before serving.

Dippers and ladles are helpful in portion control. To obtain uniformity, portion control, and cost control, it is necessary to standardize the amounts of ingredients. Here are guidelines to ensure portion control.

- Specify the number of portions to be made from large fruits or vegetables, such as number of wedges from a tomato.
- Specify the number of pieces per serving for items, such as slices of hard-cooked eggs, tomato slices, carrot sticks, or green pepper rings.
- Specify the size of the dipper or ladle to measure salad mixtures, such as potato, tuna fish, or cottage cheese.
- Specify the number of portions to be cut per pan of molded salad.
- Have available in the preparation area a picture of the way the salad should look on the plate, to ensure that all salads are alike and the portions are controlled (see the Apple Wedge Salad recipe, which includes a picture of the completed salad).

Apple Wedge Salad

Yield: 96 portions
Portion size: 5 apple wedges

Amount	Ingredients	Amount	Procedure
	Endive.............................	3 lb	1. Wash and chill endive.
	Apples, eating variety, red-skinned, medium size	24	2. Wash apples; do not pare. 3. Cut apples into quarters, remove cores, and cut each quarter into 3 wedges. Dip into oxidant according to package directions.
	Apples, eating variety, golden-skinned, medium size	16	
			4. Arrange fruit as illustrated. 5. Serve plain or with Honey Lime, Celery Seed, Poppyseed, or Sweet and Sour Dressing.

Source: Laboratory Directions for Hotel, Restaurant, and Institution Management, Iowa State University, 1992.

GARNISHES

Garnishes also can be a part of a salad. Well-planned salads usually require little, if any, garnishing because the color of the ingredients provides sufficient contrast. If garnishes are used, they should be selected to provide contrasts in color and texture and complement the other foods served. Here are some garnish ideas.

Type of Garnish	Suggestions
Vegetable	Shredded carrots
	Sliced or fluted cucumber
	Cherry tomatoes or tomato wedge
	Rings of green or red pepper or red onion
	Diced green pepper
	Ripe, green, or stuffed olive
	Radish rose or slices
	Marinated onion slices or tomato wedges
	Tiny new onions or pickles
Fruit	Fresh or frozen blueberries
	Kiwi or starfruit slices
	Dark sweet cherries
	Whole or stuffed prunes
Other	Shredded or grated cheese
	Sliced or chopped hard-cooked eggs
	Bacon chips
	Toasted or plain croutons
	Toasted chopped nuts
	Sunflower seeds
	Whole or stuffed dates

SALAD DRESSINGS

Salad dressing may be served with all types of salads to add flavor. There are no specific guidelines on which dressing to use for a certain type of salad, but the dressing should complement and blend with the other flavors in the salad. Dressings should be thin enough to mix with the salad ingredients easily, yet not runny enough to drain rapidly to the bottom. Salad dressings should not be used in excess. They should lightly coat the salad ingredients rather than smother them.

The flavor of salad dressings varies with the ingredients. Oil is the main ingredient in most salad dressings. The oils used in making salad dressings are peanut, cottonseed, corn, soybean, and olive. Because most of the calories of salads are in the salad dressing, the calories can be easily controlled by portioning the dressings. Low-calorie dressings can also be used to limit calories.

The flavor of salad dressing can become strong and spoil when it is stored in a warm place. All salad dressings should be stored in the refrigerator, once the container has been opened. Cooked dressings will spoil easily because they contain eggs.

Types of Salad Dressings

Salad dressing often consists of an oil, an acidic liquid, and seasonings. Dressings are an emulsion, a suspension of droplets of oil in a liquid, that can be prepared by a food processor, a blender, or a mixer on high speed. They can be temporary emulsions that have to be shaken before use, semipermanent ones that remain in emulsion for a few days, or permanent emulsions. An emulsifying agent, such as dry seasonings, egg yolk, or gelatin, works as a binder to hold the droplets of oil in suspension.

TEMPORARY EMULSION. French, Italian, and vinegar and oil dressings are examples of temporary emulsions. This type of salad dressing is prepared by shaking the mixture of oil and acid plus dry ingredients each time it is used. Oil is broken into small droplets by shaking, but these droplets are not small enough to stay separate in the acid. They form an oil layer when allowed to stand. The dry seasoning ingredients in French dressing work as an emulsifying agent to keep the oil in droplets.

SEMIPERMANENT EMULSION. A semipermanent emulsion is unstable, but the ingredients do not separate from each other when allowed to stand only for a short time. A thick liquid ingredient such as honey, cooked sugar syrup, or starch-thickened sauce can be an emulsifying agent. Emulsifying agents are used to increase thickness and stability. Most sweet dressings for fruit salads and herb dressings are included in this group. When these dressings separate, they can be recombined by stirring, whipping, or shaking.

PERMANENT EMULSION. Mayonnaise is an example of a permanent oil-in-water emulsion. Pasteurized egg yolk is the emulsifying agent. Because egg yolk is a combination of oil and water, when it coats the oil droplet, it aids in keeping the oil permanently suspended in the water.

Starch is used to thicken salad dressings that resemble mayonnaise. Because egg yolk is not necessary in these dressings, the fat and cholesterol content can be lowered.

COOKED SALAD DRESSING. A cooked salad dressing consists of cooked white sauce, eggs, and vinegar or lemon juice. It is fluffy and creamy with a tart flavor. Sweet cooked dressings are often used for fruit salads.

Selecting Salad Dressings

Selection of the right dressing for a salad is important in order to have the best flavor and to have a salad that goes well with the rest of the menu. Here are general suggestions in selecting the dressing:

- Salads with a strong flavored body need a mild-flavored dressing. Example: Tuna fish salad with mayonnaise.
- Salads with a mild-flavored body need a highly seasoned dressing. Example: Lettuce salad with a strong, sharp flavor, such as Italian dressing.
- Sweet salads need a sweet dressing. Example: Fruit salad with a sweet dressing, honey, or lime juice.
- The color and appearance of the dressing should make the salad more attractive. Example: Mixed green salad with red French dressing.

SUMMARY

Salads add variety, texture, color, and refreshing flavor to meals. High-quality ingredients and proper methods for preparing, storing, and serving are essential. The four basic parts of salad are the base or liner, the body, the dressing, and the garnish. Salad ingredients may include almost any raw, cooked, canned, or frozen vegetables and fruits, as well as protein foods.

Salad dressing often consists of an oil, an acidic liquid, and seasoning. The dressing should complement and blend with the other flavors in the salad. It should lightly coat salad ingredients.

SALADS 67

LEARNING ACTIVITIES

Activity 1: Preparing Tossed Salad

1. In a quantity foodservice laboratory or in institutional setting, prepare the Tossed Salad recipe for 50 portions and the Italian Dressing recipe for a half gallon.

2. Divide the recipe into two equal batches. If a class setting is not available for the suggested 2-day activity, the instructor can complete the Day 1 procedure before class. The Day 2 procedure continues as specified.

3. On Day 1 do the following steps:

    A. Wash, drain, and chill one-half the recipe. In addition, use 2 or more of the contrasting vegetables from the optional garnishing list to provide the necessary eye appeal.

    B. Store the ingredients in a covered plastic container, pan, or approved plastic bag and refrigerate for 24 hours. Avoid tightly packing or sealing the product in a container without adequate air space.

4. On Day 2 do the following steps:

    A. Prepare the second half of the recipe with prepared salad greens.

    B. Place each salad in a separate bowl and place the two bowls beside each other. Fill in the chart to compare the two products.

	Italian Dressing		
Yield: 1 gal			
Amount	Ingredients	Amount	Procedure
....	Mustard, dry	2 TBSP 2 tsp	1. Combine dry ingredients in 5-qt mixer (with splash cover) or in 12-qt mixer.
....	Salt	½ c	2. Mix thoroughly.
....	Pepper, black	1 TBSP	
....	Oregano, ground	2 tsp to 1 TBSP	
....	Garlic salt	1 TBSP 1 tsp	
....	Onion salt......................	2 TBSP	
....	Vinegar	1¼ qt	3. Add vinegar to dry ingredients; mix with flat beater until salt is dissolved.
....	Oil, salad	2¾ qt	4. Add salad oil and beat until well blended.
			5. Chill.
			Note: This dressing separates upon standing, so beat well before adding to salad ingredients.

Source: *Standardized Quantity Recipe File* (1971).

Tossed Salad

.... Yield: 100 portions
.... Portion size: 1 cup *or* 2 oz

Amount	Ingredients	Amount	Procedure
			1. Trim vegetables, discarding bruised parts, stems, etc. 2. Wash; drain thoroughly. 3. Store covered in refrigerator for several hours or overnight.

....
....
....
 	
....
	Lettuce, coarsely cut ....
	Endive, coarsely cut ....
	Romaine, coarsely cut ....
	Total weight
	Dressing: oil and vinegar,
	French or Italian, approx ....
	
	For variety and color contrast, one or more of the following vegetables may be added, substituting for an equal volume of greens:
	Radishes, sliced
	Cabbage, red, shredded
	Carrots, coarsely shredded
	Green pepper, diced or strips
	Tomato wedges
		7 lb
		1 lb 8 oz
		4 lb
		12 lb 8 oz
		
		1½ qt
			4. Cut or tear chilled greens into bite sizes (approx 1").
			5. Combine greens in chilled bowl on ring stand.
			6. Toss lightly, cover, and refrigerate.
			7. To serve:
			  a. Portion salad with tongs into chilled bowls; add dressing just before serving.
			    or
			  b. Just before serving, add dressing, toss lightly, and portion with tongs into chilled bowls.
			    Note: Toss chilled contrasting color vegetables (except tomato wedges) with greens before adding dressing; add tomato wedges just as salad is served.

Purchasing Guide

Food as Purchased	For 100 Portions	For ____ Portions
Head lettuce, approx	10 lb (5–6 heads)	
Endive, approx	3 lb (3–4 bunches)	
Romaine, approx	5–6 lb (7–8 bunches)	
When available, celery cabbage, spinach, or escarole may be substituted for an equal weight of one of the greens.		

Source: *Standardized Quantity Recipe File* (1971).

SALADS

Activity 1, Step 4B

	Salad Chilled 24 Hours	Freshly Prepared Salad
Appearance		
Texture and crispness		
Eye appeal		
Taste		

Activity 2: Preparing a Vegetable or Fruit Salad

1. Select the salad you want to prepare from the following descriptions of a fruit salad and a vegetable salad.

VEGETABLE SALADS. When using vegetables for salads, be sure to choose vegetables that are not already in your menu. The colors and flavors must blend with the rest of the meal. Here are suggested combinations for vegetable salads.

Beets
- julienne, with celery
- sliced pickled, with sliced eggs
- cubed, with cucumbers, pickle relish, and onions
- ground, with ground cabbage, onions, and horseradish, in lemon gelatin

Broccoli
- marinated, with tomato sections

Cabbage
- with pineapple and marshmallows
- with oranges and celery seeds
- with carrots, green peppers, and nuts or raisins
- with cauliflower, pickles, olives, and relishes, in lemon gelatin
- with tomatoes and cucumbers
- with red cabbage and peanuts
- with unpeeled apples

Carrots
- with sliced tomatoes, celery, and apples
- shredded, with pineapple and raisins
- in fine strips, with strips of celery, cucumber, and green peppers
- in sticks on the relish plate, with spiced crab apples and celery curls
- with cooked peas, cooked cauliflower, olives, onions, and radishes
- with raisins
- shredded, with diced apples
- with cauliflower and radishes

Green beans
- frenched, with onion and bacon
- with chopped green onions and sliced radishes
- with ripe olives and pimento
- with shredded cabbage, diced celery, onions, and green pepper

Peas
- with pickles and cheese
- with pickles, peanuts, and pimento

Spinach
- chopped with julienne beets, in sweet-sour dressing
- with bacon dressing

Sprouts, bean and alfalfa
- added to vegetable salad

Tomatoes
- wedges, with head lettuce and hard-cooked egg
- with chicory and scored cucumber slices
- cut in sixths almost through, spread open, and place a scoop of cottage cheese, coleslaw, potato salad, or egg salad in center

Vegetable to marinate
- carrots
- peas
- cauliflower
- asparagus
- tomato wedges
- green beans
- beet slices
- celery

FRUIT SALADS. Fruits are not only an appetizing and attractive addition to your menus but are also rich sources of vitamins and minerals. Here are suggested combinations for fruit salads.

Apples
- with bananas and pineapple
- with celery, oranges, nuts, and grapes
- with avocados and watercress
- with oranges, dates, and cheese
- with celery, raisins, and grapes
- with dates or raisins and peanut butter dressing
- cinnamon or spiced, filled with cottage cheese and nuts

Apricots, peach half, or pear half
- with cream cheese, cottage cheese, or cheddar cheese, and Tokay grapes
- in Waldorf salad (diced apples, chopped walnuts, mayonnaise, and other ingredients as desired)
- with blueberries
- with figs or prunes stuffed with cheese and nuts

Bananas
- with bing cherries, nuts, and cheese balls
- with nuts and celery
- with strawberries and pineapple
- with peaches, grapes, and nuts
- with apricots
- with oranges, strawberries, and pineapple
- pieces rolled in dressing, then in chopped nuts or cornflakes, and arranged with tangerines

SALADS

Cantaloupe
- with berries, pineapple, and cherries

Grapefruit
- with avocado slices
- with orange sections and green pepper slices
- with peaches, bananas, cottage cheese, and strawberries
- with a prune stuffed with cream cheese
- with orange sections, apple wedges, peaches, pears, and bananas

Oranges
- with nuts or coconut and a cherry
- with cranberries, oranges, and nuts, in cherry gelatin
- with grapes
- with Bermuda onion rings and cream cheese balls
- with mincemeat
- with coconut and pineapple
- with a red apple fan
- with grapes, bananas, nuts, and marshmallows

Mandarin oranges
- with prunes
- with cottage cheese, nuts or coconut, and pineapple chunks
- with grapefruit, in red or green gelatin
- with grapefruit sections and green grapes

    A. Assemble all the ingredients, cutting boards, a sharp stainless steel knife, and storage containers before starting.

    B. Organize your work area to minimize body movements. For example, locate a ring-stand bowl or large stainless steel container next to your cutting board to enable you to slide the cut ingredients directly into the mixing bowl.

    C. Keep the work-surface height comfortable to allow for easy circular motions of the arms and hands. No excessive reaching or bending should be necessary.

    D. Plan your cutting and mixing procedures to minimize repetitive movements. For example, rather than slice each apple cube separately, cut 3 or 4 sliced apple sections at the same time.

2. Describe a salad you have made that was well received by the people in your facility. Fill it in on the blank recipe form on page 72 and draw a picture of it.

Activity 2, Step 2

Product:		Pan Size:	
Portions:			
Portion Size:			
Ingredients	Amount	Procedure	
Picture:			

Activity 3: Comparing Storage Procedures for Lettuce

1. Demonstrate how storage procedures affect the hydration or dehydration of lettuce.

    A. Cut out the core of a fresh head of lettuce with a sharp, pointed knife.

    B. Run cold tap water into the cut end and loosen the leaves.

    C. Place the lettuce, cut end down, on a wire rack to let the water drain out.

    D. Separate the leaves and store one leaf in each way listed in the chart.

2. Record the appearance of each leaf after 24 hours.

Activity 3, Step 2

Method of Storage	Appearance after Treatment (crisp, fairly crisp, wilted, no discoloration, dried and discolored, slightly discolored)
Room temperature, unwrapped	
Refrigerator, unwrapped	
Airtight jar at room temperature	
Airtight jar in refrigerator	

SALADS

REVIEW QUESTIONS

True or False

1. Salad greens should be washed and drained thoroughly and then refrigerated until served.

   A. True
   B. False

2. To prepare a molded salad with unflavored gelatin, the gelatin needs to be soaked in hot water first.

   A. True
   B. False

3. Fresh pineapple or kiwi can be used in making gelatin salad.

   A. True
   B. False

4. Frozen fruit can be used in making gelatin salad.

   A. True
   B. False

5. When unmolding gelatin salad, dipping the mold in hot water just to the rim for a few seconds makes it release easily.

   A. True
   B. False

6. French dressing is an example of temporary emulsion.

   A. True
   B. False

Multiple Choice

7. Which of the following makes it necessary to increase the amount of gelatin in a product?

   A. Using milk as liquid
   B. Whipping gelatin
   C. Addition of a small amount of fruit

8. When is the best time to add salad dressing to salad greens?

   A. Right after portioning greens on the plate
   B. As soon as the dressing is ready
   C. Right before serving

9. Which of the following is *not* true for vegetable preparation?

   A. Salad greens should be washed, drained thoroughly, and chilled.
   B. Salad greens can be cut with any sharp knife.
   C. If salad greens are cut into very small pieces, they will wilt and lose nutrients easily.

10. Which of following is correct in preparing protein salad?

    A. The cutting board and knife should be sanitized only after finishing the salad preparation.
    B. Fully cooked eggs can be kept at room temperature.
    C. The protein food ingredients should be refrigerated, because they spoil easily.

11. When serving salads, how should the salad dressing be added?

    A. Spooned generously over salad ingredients
    B. Portioned carefully in order not to smother salad ingredients
    C. Already mixed with salad ingredients

12. Which of the following is not true of mayonnaise?

    A. It is a permanent oil-in-water emulsion.
    B. It consists of egg yolk, vinegar, seasonings, and oil.
    C. It is a semipermanent emulsion.

Matching

13. Match the salad term in column A with the definition in column B by writing the correct letter from column B in each blank.

Column A	Column B
_____ 1. Base of salad	A. A mixture of oil and vinegar or mayonnaise and seasonings
_____ 2. Body of salad	B. An emulsion of egg, oil, and vinegar
_____ 3. Molded salad	C. A leaf of lettuce or cabbage on which the other ingredients are placed
_____ 4. Accompaniment	D. The main part of a salad, the salad filling
_____ 5. Salad dressing	E. A salad shaped in a decorative mold, such as potato, gelatin, and rice salads
_____ 6. Appetizer salad	F. A salad served with the entree
_____ 7. French dressing	G. A small, light salad served at the beginning of a meal
	H. A temporary emulsion
	I. A permanent oil-in-water emulsion

5. STARCHES, SAUCES, SOUPS, CEREALS, AND PASTAS

Starch, the complex carbohydrate that plants store through photosynthesis, is a major component in many quantity foodservice recipes. Starches are used to achieve the desired viscosity or thickness in sauces, gravies, pie fillings, puddings, and some soups. They have increased in importance in the American diet with the current trend of decreased fat and sugar and increased complex carbohydrates. This increased viscosity is the result of gelatinization of the starch. Cereal starches are the most commonly used. They include cornstarch, wheat flour, rice flour, and root starches.

STARCH-THICKENED FOODS

There are three methods to ensure a smooth, even dispersion of the starch as a thickening agent in quantity recipes.

1. *The dry method.* Mixing the starch with a large amount of another dry recipe ingredient prior to the addition of the starch to the hot liquid. Example: the Cream Pie recipe.

2. *The slurry method.* Mixing the starch with cold water to make a thin paste. Example: the Cherry Cobbler recipe for 100 portions.

3. *The roux method.* Mixing the starch with melted fat to produce a paste (commonly referred to as a *roux*). Examples: the Tomato Sauce recipe, meat gravies, cream soups.

Cream Pie

Yield: 96 portions *or* 12 9" pies
Portion size: 8 per pie

Amount	Ingredients	Amount	Procedure
....	Pie crusts, 9", baked	12	1. Make pastry; bake crusts.
....	Milk, whole	1 gal 2 qt 2¼ c	2. Heat milk to scalding temperature (185 F) in steam-jacketed kettle.
....	Cornstarch .... Sugar, granulated .... Salt	1 lb 1¼ oz 2 lb 14 oz 1 TBSP	3. Sift cornstarch. 4. Combine cornstarch, sugar, and salt in a bowl; mix thoroughly. 5. Add combined dry ingredients all at one time to hot milk while stirring; cook with steam low, until mixture thickens (185 F). *Approx time:* 4 min
....	Egg yolks	1 lb 7 oz	6. Beat egg yolks with a wire whip until well mixed. 7. Add approx 1½ qt of hot mixture to egg yolks while stirring. 8. Add egg-yolk mixture to remainder of filling in steam kettle while stirring. 9. Cook to 180 F, with steam low. *Approx time:* 1–4 min 10. Immediately pour filling into a bowl.
....	Butter, softened .... Vanilla, pure .... Total weight	6½ oz 3 TBSP ½ tsp 18 lb 8 oz	11. Add butter and vanilla, mix thoroughly with a wire whip. 12. Scale hot filling into baked, cooled, crusts. *Amt per pie:* 1 lb 8 oz 13. Immediately refrigerate. Cool just to a temperature of 100 F. *Approx time:* 12–15 min
....	Meringue (Use any standard recipe.)	4 lb 9 oz	14. Add meringue to each pie. *Amt per pie:* 6 oz *Note:* If desired, meringue may be omitted and pie topped with whipped cream or topping. 15. Refrigerate pies before cutting. *Approx time:* 20–25 min

Source: *Standardized Quantity Recipe File* (1971).

The *dry method* of adding the starch to the other dry ingredients is often used in puddings and cream pie fillings. The starch is mixed with the sugar to separate the particles of starch, call *granules*.

In the *slurry method*, cornstarch is usually added to cold water to separate the granules of starch. Starch mixed with cold water will be dispersed but will not remain in suspension more than a few minutes. The starch will settle to the bottom. This mixture, sometimes called a *slurry*, can be added to the other ingredients and then heated. The expansion or gelatinization of starch occurs when heat

Cherry Cobbler

Yield: 100 portions
Portion size: 3–3¼ oz cherries *or* #10 dipper per 1⅓ oz biscuit

Amount	Ingredients	Amount	Procedure
	Shortcake Biscuit dough	8 lb 13 oz	1. Make one recipe of Shortcake Biscuit dough and bake.
	Cherries, red, tart	18 lb 8 oz	2. Defrost cherries and drain promptly, reserving juice. 3. Put cherry juice into steam-jacketed kettle. 4. Heat to simmering.
	Sugar, granulated Salt Cornstarch Water, cool Approx weight of cooked cherry mixture	1 lb 5 oz ¾ tsp 8½ oz 2 c 21 lb	5. Add sugar and salt to cherry juice. 6. Combine cornstarch and water; stir to form smooth paste. 7. Add cornstarch-water mixture to hot cherry juice, stirring constantly. 8. Cook while stirring, until thickened and clear. *Approx time:* 15–20 min 9. Turn off heat; add cherries. 10. Scale warm cherry mixture into serving dishes. 11. Top cherries with a warm shortcake biscuit and serve immediately.

Source: Adapted from *Standardized Quantity Recipe File* (1971).

is added to the liquid ingredients. If starch granules are heated without being dispersed, a lumpy product will result. The cold slurry method also works well with flour. Fruit fillings, like the one for the cherry cobbler, usually use this method.

The *roux method* is the most frequently used method for flour-thickened sauces and gravies. Most quantity foodservices make a roux mixture for cream soups, sauces, and gravies. The fat must not be too hot when the flour is added, or the starch granules will shrink and clump together to cause lumps. A lumpy roux may also occur if the flour is added too rapidly or not stirred enough. The roux is usually sautéed until all the flour is coated with fat and the roux is light brown. The roux and liquid in the sauce or gravy are mixed together and brought to a boil. Tomato sauce made in quantity is most consistently of good quality when the roux method is used.

In a roux, starch particles, called *granules,* must be heated in liquid or subjected to moist heat for *gelatinization* to occur. Gelatinization is the process of the starch granules absorbing water and swelling. Maximum viscosity and loss of the raw starch flavor happen at the same time. The temperature needed for this stage varies with different starches, but all cereal starches used in quantity food production reach maximum thickness by the time they come to a boil. Usually flavor is improved after gelatinization is complete in cereal starches.

In all of the thickening methods, the mixtures are gently stirred during the heating of the ingredients. This helps to keep the starch particles dispersed while exposing all starch granules equally to the heat and water. Quantity production recipes often specify using a large wire whip to ensure even mixing of the starch granules. Overstirring, however, can actually break the gelatinized starch granules, resulting in a thinning of the mixture.

	Tomato Sauce		
Yield: 96 portions			
Portion size: 1-oz ladle			
Amount	Ingredients	Amount	Procedure
....	Tomato juice	3 qt	1. Pour tomato juice into steam-jacketed kettle or saucepan.
....	Onions, dehydrated	¾ oz	2. Add rehydrated onions, sugar, salt, pepper, Worcestershire sauce, and celery seed.
....	Sugar, granulated	1 oz	
....	Salt	1 tsp	
....	Pepper, black	¼ tsp	
....	Worcestershire sauce	1½ tsp	3. Heat to boiling. Reduce heat and simmer for 10 min.
....	Celery seed	½ tsp	4. Strain.
....	Tomato paste	⅔ c	5. Return strained juice to kettle or saucepan. Add tomato paste and blend.
....	Margarine or butter	5½ oz	6. Prepare roux:
....	Flour, all-purpose	4 oz	a. Melt margarine or butter.
			b. Add flour and mix until blended.
	Total weight	6 lb 3 oz	7. Heat tomato juice to boiling; add roux while stirring.
			8. Cook until thickened.

Source: *Standardized Quantity Recipe File* (1971).

Puddings and Pie Fillings

The dry method, adding the starch to the other dry ingredients, is often used in puddings and cream pie fillings. The other dry ingredients, especially sugar, help to keep the individual starch granules separate. This prevents formation of lumps after the liquid ingredients are added later and heated.

Pie fillings and pudding mixes with starches that have been cooked and dried by the manufacturer do not require heat during the stirring. This reduces preparation time for quantity foodservices.

If traditional pudding recipes are used, the quantity foodservice manager should check the recipe procedures to be sure the eggs are *tempered* (added to a small amount of the hot starch mixture to increase their temperature). The warmed eggs are then added to the hot liquid. If eggs are added directly, the egg proteins will coagulate before being blended evenly throughout the mixture, causing lumps or a curdled appearance.

Cornstarch, flour, and tapioca are starches used to thicken puddings and pie fillings. It takes twice as much flour to produce the same amount of thickening as cornstarch. Tapioca, a root starch, can be used to thicken puddings, but its thickening power is less than cornstarch. Tapioca does not form a gel when cooled, and overcooked tapioca puddings have a tendency to become gummy.

The effect of acid and sugar on starch must be considered when preparing fruit pie fillings. When preparing a filling in large volume, the acid and starch mixture must be cooled quickly in order to minimize the thinning effect the acid has on the starch. Quick cooling will stop the cooking before the acid affects the thickness. In many desserts containing fruit and juice, the acid reactions cannot be avoided. A good product can still be made by heating slowly, not overheating, and stirring carefully.

Sugar also has an effect on puddings and pie filling thickness. Too much sugar has a tendency to produce a thick syrup rather than the desired gel. The starch must compete with the sugar for the available water. Because the sugar absorbs water more easily than starch, it prevents or blocks the starch from absorbing as much water. Preventing the granules from swelling as much can, however, protect from the thinning that can result from overstirring or boiling too hard.

Although starch gels in puddings and pie fillings appear to be stable when frozen, part of the starch structure contracts. In quantity operations, using frozen pie fillings or thickened sauces can be a problem. When they are thawed, the water that was forced out of the starch during freezing will run out of the product. This *syneresis,* or loss of water, cannot be corrected by reheating. Because of these problems, chemically and physically modified starches have been developed for use in frozen foods.

Sauces

The type of starch used to produce thickened sauces varies with its use. For example, cornstarch produces a translucent sauce for sweet and sour pork and in fruit pie fillings. Cornstarch is a cereal starch with greater thickening power than root starches like potato and tapioca. In quantity foodservice operations, recipes will specify that cornstarch be added to water before being added to the sauce.

Opaque sauces and gravies usually specify flour for thickening. This will not produce a clear sauce. To save time and speed up production, a roux can be made in larger batches once a day for use in these sauces and gravies, as needed.

Dextrinization, the breaking down of the starch that occurs when flour has been browned for use in a brown sauce, reduces the thickening power of the starch. The addition of unbrowned flour to the sauce will help achieve the desired thickness.

Sauces vary in thickness according to use. They can be thin, medium, thick, or very thick. Here are uses for each type.

Sauce	Use
Thin	Cream soups
Medium	Au gratin foods, gravy, creamed vegetables
Thick	Soufflés
Very thick	Premolded fried foods, like croquettes

In white sauce, the roux is added to the liquid after it has cooked long enough to eliminate the starchy taste. This fat-starch mixture is combined with cold liquid from the recipe to disperse the starch more completely. In quantity recipes, about one-fourth of the milk is added cold to the fat-starch mixture. The remaining milk is then preheated to 160–185 F before being added. (Preheating helps to prevent lumps while also saving production time and thinning from long cooking.)

In another roux method, the roux is prepared in a steam-jacketed kettle. The cold liquid is added to the roux while stirring. With this method, the sauce will take longer to thicken after the roux and liquid are combined.

Lumping is one of the major problems in producing a quality white sauce. The sauce should be perfectly smooth. Lumps usually result from insufficient mixing of the starch with the fat or other ingredients in the recipe prior to the gelatinization of the starch. If the starch is added to fat that is too hot, lumps are likely to occur. If the sauce becomes lumpy during the initial stages of production, heat reduction and additional stirring may help to eliminate some of the lumps.

Another problem is a film of fat that sometimes appears on the surface of the white sauce. This can be the result of poorly gelatinized starch or a mixture that has too much fat. This is especially common in gravies when production personnel do not measure the meat drippings added to the gravy (see below). Gravy may be improved by reheating to boiling to gelatinize the starch completely or by adding more starch dissolved in water.

Gravies

Gravies should be smooth and free from any lumps. Because they are similar to a medium white sauce, the same principles apply to gravy production as they do to a basic white sauce. The fat or oil in the fat-starch combination is usually taken from fat drippings from roasted or fried meats. These drippings must be accurately measured to determine the amount of starch needed to thicken the gravy. It usually requires 2 tablespoons of fat and 1 tablespoon of flour to make 1 cup of gravy, based on the medium white sauce recipe.

Meats that have been cooked by the braising or stewing methods already have some liquid in the drippings. This moisture will cause immediate gelatinization of the starch if dry starch is added to the hot liquid, and lumps will form. A starch and cold-water slurry should be added to the meat juice to thicken without forming lumps.

Cream Soups

Cream soups, like cream of chicken, mushroom, and asparagus, are made using a thin white sauce as the base. (Cream of potato soup is not like the other cream soups because the thickening is partly due to the potato starch in the soup.) Cream soups can be made by pureeing all the ingredients or by adding chopped meat and vegetables.

Caution must be used in quantity production to ensure that the proteins in cream soups do not curdle. If a soup, such as cream of tomato soup, is too acidic, the milk proteins will curdle. Acidic soups must have the acid-containing foods added very slowly to the white sauce mixture to keep the milk from curdling. It is very important that the acid ingredients are added to the milk, not the reverse. Because of the problems in making the soups from scratch, most quantity foodservices use canned tomato soup.

The heating period of acid-containing foods also should be kept to a minimum. Long periods of holding cream soups under high heat will result in the soup curdling. Therefore, in quantity foodservices it is best to prepare cream soups in small batches to prevent the soup from extended exposure to a heated cafeteria or food assembly line.

CEREALS

Cereals are an important part of the diet in many countries around the world. In America, the recent emphasis on increasing the complex carbohydrates and decreasing the sugar and fat in the diet has encouraged clients to include more cereal in their diet. The favorite cereals are made from wheat, corn, rice, and oats. Barley, buckwheat, and triticale are not as popular.

Cereals are primarily made of starch. The small amount of protein in cereal is an incomplete protein (lacking some essential protein components, amino acids). The usual combination of cereal with milk greatly improves the cereal's nutritional value as it provides not only complete proteins but also provides the essential amino acids missing in the cereal. The grains from which cereals are made are excellent sources of niacin, thiamine, riboflavin, pyridoxine, and pantothenic acid. When the outer covering of the grain is included in the cereal, it is a major source of fiber.

In producing breakfast cereals, the grain is processed by polishing, rolling, puffing, granulating, or shredding. When a process removes the bran layer of the grain, most of the natural vitamins and fiber in the cereal are discarded. Some cereals are enriched to replace these lost nutrients. Many states require that thiamine, riboflavin, niacin, and iron be added to all cereals that have been refined by removing the bran layer and grinding during the milling process. Package labeling includes information on the fortification and enrichment added. *Fortification* replaces nutrients that were lost during processing, such as the addition of thiamine to wheat cereals. *Enrichment* adds nutrients to the product that were not originally in the food, as in the addition of calcium to orange juice.

Wheat is used to make many cereals. Farina is the most popular cooked wheat cereal. *Farina* is wheat that has been refined to remove the outer bran layer. *Bulgar* is the whole wheat kernel that has been precooked and cracked before it is dried and sold. *Kasha* is the cracked whole buckwheat kernel.

Corn is popular as a ready-to-eat cereal. Corn made into hominy grits is frequently served as a cooked cereal for breakfast in the southern United States.

Cooking Methods for Cereals

Because cereals are mostly starch, cereal cooking methods are similar to those using starch to thicken soups and sauces. Cereals should be boiled or cooked to at least 195 F to obtain maximum swelling of the starch and remove the raw flavor. Undercooked cereals will taste similar to raw flour and be very difficult to digest.

Each type of cooked cereal should be prepared according to a specific standardized recipe. The amount of water and cooking time will vary with the type of cereal. In addition, the length of holding time after cooking also must be considered in preparing the cereal. However, there are principles that apply to preparing all types of cooked cereal to produce the highest possible quality.

Preventing lumps:
- Fine-grain cereals should be mixed with cold water before being added to the boiling water. This is similar to the method for making a slurry of cornstarch and water for sauces.
- Cereals should be slowly added to the water while stirring.
- Adding a small amount of butter, margarine, or oil will prevent boiling over and lumping.

Cooking temperature:
- The cooking water should be boiling hard before the cereal is added.
- The cooking temperature should be high enough so it keeps boiling or simmering when the cereal is added.

Cooking time:
- The cooking time should be longer for coarse grains than for fine grains.
- Cereals must cook longer than pure starches, such as cornstarch, in order to soften the fiber and allow absorption of water.

Cooking procedures:
- Stirring too fast or hard will break up the cereal and cause it to be sticky, gummy, and too smooth.
- Cereals should be cooked in a steam-jacketed kettle, heavy pan on the range, compartment steamer, or double boiler. Single servings can be cooked in the microwave.
- Cereal should be cooked as close to the time of service as possible because it should be served hot (180 F) and because it will continue to thicken and get sticky if it is held.

Adding salt:
- The salt should be added to the cooking water before the cereal. This ensures the salt will be mixed evenly through the cereal.
- Because the salt slows starch gelatinization, salt-free cereal will take a little less time to cook.

Holding cooked cereal:
- To hold cereal after cooking, it should be covered tightly to prevent a dry skin from forming on top. If a firm skin forms, it should be removed, not stirred into the pan of cereal, because the skin may accidently be served.

Adding variety to cereals:
- Dried fruit, raisins, prunes, peaches, apricots, or canned fruit can be added to cooked cereal. Dried fruit is added early in the cooking, canned fruits just before the cereal is done.
- Dried or fresh milk can be added to cereal to increase nutritional value. More cooking water may be needed when dry milk is added. Dried milk is mixed with the dry cereal before cooking. Fresh milk can be used in place of part of the cooking water.
- Fruit juice may be added to the cooking water. The acid in the juice will cause the cereal to take longer to cook.

RICE

Rice is the usual accompaniment to Oriental and Creole entrees. As complex carbohydrates and whole grains increase in popularity, the use of rice with other foods is becoming more acceptable to clients. Rice should always be fluffy, tender, and dry. Each grain should be slightly firm but not hard or crunchy. Short- and medium-grain kernels should be sticky enough to just cling together. These types of rice are typical for authentic Oriental meals. American foods are usually served with the less sticky long-grain rice.

Long-grain and converted rice separate into individual grains. All types of rice should easily spread out on the plate when served. If the rice is very dense and stays in the shape of the dipper, it is too sticky.

Converted rice has been parboiled to force the nutrients in the outer hull into the rice grain, so it does not lose as much nutritional value as other rice-polishing methods. Converted rice has larger rice grains and is less sticky than long-grain rice but is not as soft when cooked. This rice holds very well, which is useful for quantity foodservices. However, converted rice is usually more expensive per pound as purchased than is long-grain rice.

Precooked rice has been cooked until the starch is gelatinized, then cooled and dried. The individual rice grains are not as sticky as regular long-grain rice, but they have irregular shapes and are smaller. Because it is partially cooked, precooked rice can usually be prepared in less than 10 minutes. Both converted and precooked rice have a lower yield than short-, medium-, and long-grain varieties that have not received these treatments (see Table 5.1).

Table 5.1. Equivalent amounts of three forms of rice

Type of Rice	Weight, Uncooked	Amount of Water	Weight	Cooked Measure
Long-grain	1 lb	1½ qt	5 lb	2 qt 1 c
Converted	1 lb	1 qt	3 lb	1 qt 2⅔ c
Precooked	1 lb	1 qt	3 lb	Approx 1 qt

Cooking Methods for Rice

In quantity production, rice can be cooked by steaming, simmering, or baking. The easiest method is steaming in a compartment steamer. The constant temperature, pressure, and moisture help produce a consistent product. The length of steaming may vary slightly from one steamer to another. As the pressure increases, the time will decrease. Converted rice should be steamed approximately 22–27 minutes. Long-grain rice may take slightly longer, 25–30 minutes.

All types of rice can be cooked in a steam-jacketed kettle or in a heavy pan on the range top. The rice and salt are added to boiling water, the heat is adjusted to keep the water at a slow boil, and the kettle or pan is covered. It is cooked without stirring 20–25 minutes.

Oven-cooking rice can produce a high-quality product. Boiling water is poured into a steam-table pan, and salt and rice are stirred into the water to distribute them uniformly. After covering tightly, the rice is baked. See the tested recipe for Oven-cooked Rice.

Oven-cooked Rice

Yield: 40 portions
1 12×20×2½" pan
Portion size: approx 2 oz *or* #12 dipper

Cooking temperature: 350 F
Baking time: 45 min

Amount	Ingredients	Amount	Procedure
	Water, hot	2 qt	1. Measure hot water into pan.
	Rice, converted	2 lb	2. Add rice and salt to hot water.
	Salt	2 tsp	3. Stir to distribute uniformly.
			4. Cover tightly.
			5. Bake until rice is tender and all water is absorbed.
			Note: Rice may be cooked in a steamer if desired. Cook uncovered until rice is tender and all water absorbed (approx time: 22–30 min).

Source: *Standardized Quantity Recipe File* (1971).

If rice must be held before service, a small amount of butter or margarine will keep it from getting sticky. Rice dries out quickly when held hot. Covering it tightly will help give a fluffy, tender product. When too much water is used in cooking or added during holding, the rice is stickier and cooks together into a pasty ball.

PASTA

Pasta has become much more popular in the last five years. It is a frequent menu item in nursing facilities, hospitals, and schools, as well as at home. Pasta can be used in many ways: as the main ingredient in many casseroles, added to soups, made into salads, and as the base for Italian entrees. Enriched pasta contributes complex carbohydrates, calories, and B vitamins to the diet. Over five hundred shapes and forms of pasta are available. The most common ones are elbow macaroni, spaghetti, and noodles.

Pasta is also known as *alimentary paste*. It is classified into three types, based on ingredients and shape. The flour and water products make up two types, hollow tubes, such as macaroni, and solid rods, such spaghetti. The third type of pasta is noodles, a flat ribbon shape made from egg plus flour and water.

The quality of cooked pasta is greatly influenced by the type of wheat it is made from. The highest-quality pasta contains *semolina*, which is made by milling durum wheat. The lowest-quality pasta is made from farina, a by-product of milling. Any other wheat products result in a very poor quality product. Semolina pasta is a pale straw-yellow color, has resistance to falling apart during cooking, and produces a firm yet tender cooked pasta without stickiness. A poor-quality pasta has a pale or gray color, tends to break apart or dissolve during cooking, has a soft and mushy cooked texture, and clings together.

Pasta made from scratch is becoming more popular in fine restaurants. It can be made in any quantity production facility from hard wheat flour, water, and salt. However, from-scratch products will not have the same quality characteristics as purchased dried pasta for three main reasons:

- Purchased pasta is made from a special, coarsely ground wheat product.
- Purchased pasta is drier before cooking.
- Purchased pasta is shaped by high-pressure extrusion, and production facilities then roll and cut it.

Pasta made from scratch will have a softer more doughy texture than the purchased dry types. Individual opinions vary on which of the two products is higher quality.

Noodles

Noodles can be purchased dried, precooked, or frozen, or can be made in the facility. The quality characteristics of desirable noodles are similar to other pastas. Noodles will have a softer texture than the hollow tube or solid rod pastas because they are made from a finer milled flour.

The thin ribbons of noodles easily cling together if held hot after cooking. Thicker varieties hold better than thin ones. Frozen noodles have a very desirable texture if held for a short time. They are firmer, tougher, less sticky, and thicker than dried noodles. In any cook-chill and cook-freeze food-production system where foods are prepared, stored, and reheated, the use of frozen noodles is recommended.

Cooking Methods for Pasta

Pasta can be boiled or steamed. Steaming in a compartment steamer is preferred for lasagna noodles and other large pastas. In general, steaming is the best cooking method for pasta that will be assembled into a casserole, like lasagna, or for manicotti tubes or large shells that are stuffed. The steaming helps prevent the large pasta from breaking. Because steaming takes longer and does not allow the stirring that prevents pasta from sticking together, boiling is better for the smaller pasta shapes.

The recommended boiling procedures are listed in Table 5.2. The pasta is added to rapidly boiling salted water. The heat should be adjusted to maintain a slow boil. Gentle stirring is needed to keep pasta from sticking together. A small amount of oil in the cooking water is also recommended. About 1 tablespoon of oil for each 1 pound of pasta will help prevent sticking without making the pasta greasy.

Here are characteristics of high-quality pasta when it is done:

- It has a consistent firm texture throughout each piece with no firm center core.
- It is slightly chewy, not mushy or sticky.
- It has a uniform color.
- It clings to a cold, clean, stainless steel utensil.

Pasta should be drained and rinsed as soon as it is cooked to the correct texture. The rinsing removes any extra starch that promotes stickiness and it stops the cooking process. Cool water should be used to stop the cooking, yet still leave a product hot enough to serve.

Table 5.2. Recommended cooking methods and yields for pasta

Type of Pasta	Weight, Uncooked	Cooking Method	Cooking Time	Weight, Cooked
Small hollow tubes				
Macaroni	1 lb	Boiling	5–7 min	3 lb
Small shells	1 lb	Boiling	5–6 min	2.5 lb
Solid rods				
Spaghetti	1 lb	Boiling	6–8 min	3 lb
Noodles, dried				
Lasagna	1 lb	Pressure steaming	25 min	3 lb
Lasagna	1 lb	Boiling	12–15 min	3 lb
¼" flat	1 lb	Boiling	4–5 min	3 lb
Noodles, frozen				
¼" flat	1 lb	Boiling	15 min	2 lb

Holding Pasta

There are several ways to hold pasta after cooking, but it is best to hold it only a minimum amount of time by cooking in smaller batches. Even in small batches, the last portions may stick together. To ensure that all servings are of the best quality, the following four methods are helpful.

- Mix the pasta with a sauce before holding.
- Add butter or margarine to the pasta before holding.
- Cook the pasta in small batches that will be served immediately.
- Cook the pasta in large batches and hold in cool water.

The following procedure holds the large batch of pasta in cool water and reheats small batches just before service.

1. Cook the pasta until just done.
2. Drain the pasta, then cover it with cool water. Rinse it only if it is very sticky.
3. Place small batches of the cool pasta in a strainer.
4. Immerse the strainer and pasta in boiling water for 1 minute.
5. Drain the pasta and serve immediately. Do not rinse it after reheating.

Holding the pasta in cold water is preferred because it gives the highest-quality product, without added sauce or fat.

SUMMARY

Starch products are a major part of the American diet. They include starch-thickened sauces, puddings, and soups; cereals; rice; and all types of pasta. All of the starches provide complex carbohydrates, calories, and B vitamins. Some products must be fortified to restore the vitamins after processing.

All starches cook in similar ways. The starch must absorb water when heated. As a thickening agent, the starch granules swell to form a thicker product. In cereals, rice, and pasta the absorption of water tenderizes and hydrates the product. To be digestible and have desirable flavor and texture, all starches must reach at least 195 F during cooking. Most will benefit from being cooked to near 212 F. Overcooking, however, causes a decrease in quality that usually cannot be reversed.

LEARNING ACTIVITIES

Activity 1: Comparing Cooked Pastas

1. Purchase a variety of pastas that are different in shape. Include at least small shells, macaroni, dried fettuccine noodles, and lasagna noodles.

    A. Cook 8 ounces of each pasta separately according to the instructions in Table 5.2.

    B. Compare the doneness of each pasta.

    Which pasta held together best? _____

    Did some break apart or stick together more than others? Which ones?
    _____

2. Compare holding methods.

    A. Place half of the cooked pasta in cool water and hold at room temperature 30 minutes. Use a glass or stainless steel bowl.

    B. Divide the remaining pasta and place half of it in the top of two double boilers. Place the pasta over simmering water.

    (1) Add hot water (160 F) to the pasta in one pan and cover.

    (2) Add 2 tablespoons of butter or margarine to the pasta in the other pan and cover.

    (3) Hold the pasta 30 minutes. The water in the bottom of the double boiler should be slowly simmering.

    C. At the end of the 30-minute holding time reheat the pasta held in the cold water.

    (1) Heat 2 quarts of water to a fast boil.

    (2) Place it in a strainer and drain.

    (3) Dip the strainer and pasta in the boiling water for 1 minute.

    (4) Drain and then immediately begin step D.

    D. Compare the three methods for holding pasta. Taste and look at the pasta. Evaluate their flavor, texture, and labor time.

    (1) Are there differences in the flavor among the three samples?

    (2) What are the textures of the three samples?

    Held cold _____

STARCHES, SAUCES, SOUPS, CEREALS, AND PASTAS

Held in hot water _____

Held with fat added _____

Which has the best texture? _____

Which method would work best in your facility and why?

_____

_____

Can the method you selected be used for all pasta recipes and those for diet modifications?

_____

Activity 2: Comparing Cooked Cereals

1. Using a standardized recipe or package directions, cook 2 portions of these three cereals: old-fashioned, long-cooking rolled oats; quick-cooking (5-minute type) oatmeal; and instant oatmeal packets.

2. What is the texture of each oatmeal?

   Long cooking _____

   Quick cooking _____

   Instant _____

3. Compare the texture of each oatmeal with the quality standards you have learned. Rank the three oatmeals for texture quality.

   Best _____

   Second _____

   Third _____

4. Taste each oatmeal. Notice the flavor, saltiness, and how it feels in your mouth. What are the differences among the three samples in flavor, salt, and mouth feel?

   _____

   _____

Activity 3: Comparing the Slurry and Dry-Mix Methods

1. Using the Cherry Cobbler recipe for 25 portions and the Shortcake Biscuit recipe adjusted to 50 portions, prepare one batch of cobbler using the slurry method described in the recipe's procedures.

2. Make a second batch of cobbler, following the same recipe, but substituting the following steps for 4, 5, 6, and 7. These steps use the dry-mix method.

    4. Heat the water and cherry juice to simmering.
    5. Combine the sugar, salt, and cornstarch together.
    6. Add all dry ingredients at once to the heated water and cherry juice solution.
    7. Stir the mixture until the dry ingredients are dissolved.

Cherry Cobbler

Yield: 25 portions
Portion size: 3–3¼ oz cherries *or* #10 dipper per 1⅓ oz biscuit

Amount	Ingredients	Amount	Procedure
	Shortcake Biscuit dough	2 lb 3 oz	1. Make one recipe of Shortcake Biscuit dough and bake.
	Cherries, red, tart	4 lb 2 oz	2. Defrost cherries and drain promptly, reserving juice. 3. Put cherry juice into steam-jacketed kettle. 4. Heat to simmering.
	Sugar, granulated Salt Cornstarch Water, cool Approx weight of cooked cherry mixture	5 oz ¼ tsp 2 oz ½ c 5 lb 4 oz	5. Add sugar and salt to cherry juice. 6. Combine cornstarch and water; stir to form smooth paste. 7. Add cornstarch-water mixture to hot cherry juice, stirring constantly. 8. Cook while stirring, until thickened and clear. *Approx time:* 15–20 min 9. Turn off heat; add cherries. 10. Scale warm cherry mixture into serving dishes. 11. Top cherries with a warm shortcake biscuit and serve immediately.

Source: *Standardized Quantity Recipe File* (1971).

STARCHES, SAUCES, SOUPS, CEREALS, AND PASTAS

Shortcake Biscuits			
Yield: 100 servings Portion size: 1⅓ oz or #30 dipper		Baking temperature: 375 F Baking time: 15–25 min	
Amount	**Ingredients**	**Amount**	**Procedure**
....	Flour, all-purpose .... Baking powder .... Salt .... Sugar, granulated	3 lb 4 oz 3 oz 1 TBSP 1½ tsp 1 lb	1. Combine flour, baking powder, salt, and sugar in 20-qt mixer. 2. Using pastry blender, mix on low speed until blended.
....	Butter	1 lb 10 oz	3. Cut butter into flour mixture with pastry blender; mix on low speed to a coarse, mealy consistency. *Approx time:* 2 min
....	Cream, whipping .... Water, room temperature .. Total weight	3¼ c 1½ c 8 lb 8 oz	4. Combine whipping cream and water. 5. Add liquids to flour mixture in a steady stream. Mix just enough to moisten. *Approx time:* 30 sec 6. Portion biscuits onto ungreased 18×26×1″ pan, leaving approx 2″ between biscuits. *No. per pan:* 35 7. Bake for 10 min at 375 F, then lower temperature to 350 F and bake until lightly browned. *Approx time:* 15–17 min total 8. Serve warm as a top for cobbler, or for Strawberry Shortcake.

Source: Laboratory Directions for Hotel, Restaurant, and Institution Management, Iowa State University, 1992.

3. Compare the two products.

   A. In which recipe was it easiest to dissolve the cornstarch into the liquid ingredients? Why?

   _____

   B. What happens to the starch granules when added to a very hot, simmering liquid?

   _____

   C. Which of the two recipes did you prefer? Why?

   _____

   D. Complete the comparison chart.

Activity 3, Step 3D

	Slurry Method	Dry-Mix Method
General appearance		
Consistency (viscosity) of filling		
Color of filling		
Taste		

REVIEW QUESTIONS

True or False

1. The highest-quality macaroni is made from semolina.

    A. True
    B. False

2. The highest-quality macaroni is made from durum wheat.

    A. True
    B. False

3. The yields of 1 pound of frozen noodles and 1 pound of dried noodles are both 3 pounds.

    A. True
    B. False

4. Cornstarch, flour, and tapioca are all starches that can be used to thicken puddings and pie fillings.

    A. True
    B. False

5. Opaque sauces and gravies usually specify cornstarch for thickening.

    A. True
    B. False

6. Tapioca, a root starch, has more thickening power than cereal starches.

    A. True
    B. False

STARCHES, SAUCES, SOUPS, CEREALS, AND PASTAS

Multiple Choice

7. Gelatinization is the process when starch

    A. Tightens when frozen
    B. Begins to absorb water and swell
    C. Reaches maximum thickness

8. In starch-thickened mixtures, sugar causes the product to be

    A. Thicker
    B. Thinner

9. In making a cornstarch-thickened cherry sauce, the cook should

    A. Stir fast to avoid lumps
    B. Boil the sauce for 10 minutes
    C. Hold the sauce at 190 F until serving time
    D. Stir gently to avoid breaking the starch granules

10. Lemon juice added to a cornstarch pudding at the beginning of cooking will

    A. Speed up thickening
    B. Slow down thickening

11. The most frequently used technique for flour-thickened sauces and gravies is this method.

    A. Slurry
    B. Roux
    C. Dry
    D. Liquid

12. One cup of sauce based on a medium white sauce recipe requires

    A. 1 tablespoon of fat and 1 tablespoon of flour
    B. 2 tablespoons of fat and 1 tablespoon of flour
    C. 2 tablespoons of fat and 2 tablespoons of flour
    D. ½ tablespoon of fat and 1 tablespoon of flour

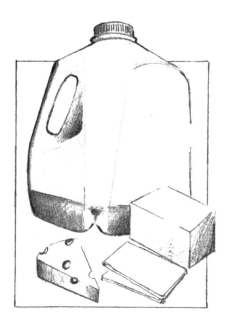

6. MILK AND CHEESE

MILK

Milk is a complex food that has important nutritional value for Americans. Milk sold in this country must be pasteurized or ultrapasteurized and cannot contain less than 8.25% milk solids. Whole milk must have at least 3.25% fat. Individual states may have regulations with higher standards than the federal regulations. Iowa regulations for milk served as a beverage are included in Chapter 10.

The amounts of fat, protein, vitamins, and minerals in milk vary with the cow's feed, health, and breed. Milk contains mostly water, about 87% of the total. Milk proteins have the most influence on its cooking properties and the manufacture of dairy products (see Table 6.1).

Nutritional Values of Milk

PROTEIN. Milk is a complete source of protein, providing all essential amino acids. The amino acid casein accounts for 80% of the protein in milk and is responsible for the curdling properties of milk and the gelling of milk when acid is added. The curdling of milk is a necessary part of making cheeses, when it separates the portion of the milk protein used for cheese from the whey. A familiar example of milk gel caused by bacteria-releasing acid is yogurt. The whey proteins are in the liquid portion of milk after the casein curds have been removed.

ENZYMES. Most enzymes in milk are destroyed by pasteurization. This destruction is used, in fact, to find out if the pasteurization process is completed successfully. When enzyme activity is no longer present, the milk has been completely pasteurized.

CARBOHYDRATES. *Lactose* is the major carbohydrate in milk. About 4.8% of the milk is lactose. This amount of lactose does not make the milk taste sweet because the flavor of lactose is less sweet than that of *sucrose* (granulated sugar). Milk also contains very small amounts of other carbohydrates: *glucose* and *galactose,* as well as other types of sugars. This information is important to

Table 6.1. Content of milk products

Milk	Water, %	Protein, %	Carbohydrate, %	Fat, %	Calcium mg/100 gm
Fresh					
Whole	88	3.3	4.7	3.34	119
Low-fat, 2%	89	3.3	4.8	1.9	122
Canned evaporated					
Whole	74	6.8	10.04	7.6	261
Skim	79	7.6	11.4	0.2	290
Sweetened condensed	27	7.9	54.4	8.7	284
Dried					
Whole	2.5	26.32	38.4	26.7	912
Nonfat	3.2	36.2	51.98	0.8	1257
Cream					
Half-and-half	80.6	2.96	4.3	11.5	105
Whipping	57.7	2.1	2.8	37.0	65
Cheese					
Cheddar	36.8	24.9	1.3	33.1	721
Cottage, creamed	79.0	12.5	2.7	4.5	60
Cottage, 2% fat	79.3	13.7	3.6	1.9	68

foodservice personnel because clients may have allergies to specific carbohydrates.

Lactose intolerance occurs when the lactose cannot be digested because an individual lacks the enzyme *lactase*, which is needed to digest lactose. Lactose-intolerant people experience discomfort when the undigested lactose gets into the large intestine. These people can drink milk treated to remove the lactose. Naturally processed cheese and yogurt can also be eaten because the lactose is converted to lactic acid in making these products.

VITAMINS. Milk contains many vitamins. *Riboflavin,* a B vitamin, is present in important quantities.

This vitamin gives skim milk its light blue-green tint. Since exposure to light destroys this vitamin, it is better to store milk in containers that protect the milk from light. Cardboard cartons and heavily frosted plastic jugs provide enough protection for riboflavin.

Milk contains *carotene,* which is changed into vitamin A by the body when milk products are eaten. Carotene is what gives milk its creamy yellow color. In low-fat and fat-free milk products, the carotene is removed from the milk along with the fat. These foods are fortified to replace the lost vitamin.

Milk and milk products are the major sources of *calcium* in the diet. Vitamin D is added to milk to help people absorb the calcium.

Milk Processing

HOMOGENIZED MILK. The process of breaking the milk fat into very small particles is called *homogenization*. The milk is forced through very small slits to break up the fat. This prevents the fat particles from joining with each other, allowing the milk fat to float to the top. Homogenization is necessary to keep the fat distributed in milk products. It causes milk to curdle faster and thicken more slowly. Homogenized milk forms firmer custards and puddings than does unhomogenized milk.

PASTEURIZED MILK. *Pasteurization* is a heat treatment of milk. It may be completed several ways, heating the milk to a specified temperature for the minimum time needed to destroy the disease-causing microorganisms and bacteria that cause spoilage. In addition, off-flavors caused by the enzyme activity are destroyed.

The flavor of milk is changed by pasteurization. Using a high temperature for only a short time changes the flavor less than does scalding or boiling milk.

Ultrapasteurization is currently used to extend the shelf life of half-and-half and whipping cream.

With ultrapasteurization and storage at temperatures below 42 F, cream can be stored for up to 6 weeks before it is opened. A temperature of 212 F for 0.01 seconds kills more bacteria than the lower temperatures used in pasteurization. Some additional flavor changes are caused by a higher temperature, but the flavor change is less noticeable in cream than in milk.

EVAPORATED MILK. Evaporated milk is made by removing some of the water from fresh fluid milk.

Canned evaporated milk has a little more than half the water removed. The process includes evaporation, homogenization, canning, and sterilizing. The usual flavor of canned milk is caused by the high temperature. During the evaporation process the milk must be held at 239 F in a partial vacuum for 15 to 20 minutes.

SWEETENED CONDENSED MILK. Sweetened condensed milk should not be confused with canned evaporated milk. Sweetened condensed milk has enough sugar added to make the total carbohydrate content 56%, compared with the 10% carbohydrate content of evaporated milk. Because the high sugar content in sweetened milk acts as a preservative, sterilization is not needed.

DRY MILK. Dry milk is available in whole milk powder, dry buttermilk, and nonfat dry milk. Whole dry milk becomes rancid more easily than the lower-fat dry buttermilk and nonfat dry milk. The largest use of dry milk is as an ingredient in commercially prepared mixes and in quantity food production. Nonfat dry milk is over 75% of the dry milk produced.

Nonfat dry milk will keep up to 18 months if stored correctly. It should be stored in a dry storage area, at a temperature below 70 F. If the milk becomes damp it will become lumpy and will be very difficult to use. When dry milk is stored above 70 F, it will have a stale flavor and a tan color. Once reconstituted, all dry milk must be stored in a refrigerator.

Cooking with Milk

The changes that happen to milk in cooking are similar to those in pasteurization. The heat affects the flavor and texture. When milk is heated to a temperature above 185 F and held at that temperature for more than a few seconds, the proteins begin to cook. This cooking causes the milk to form a thin layer of cooked whey at the bottom of the pan, which easily scorches and burns, giving the milk a burnt flavor. This layer at the bottom of the pan can turn brown even at temperatures below 185 F.

The use of a compartment steamer or steam-jacketed kettle to heat milk provides good temperature control. In a compartment steamer the whey that settles to the bottom will not scorch because the bottom of the pan is not in contact with a heat source above 240 F. The steam-jacketed kettle helps the cook use temperatures below 240 F to heat the milk. However, the layer of whey will scorch if the steam is not controlled. Milk easily foams and boils over the sides of the kettle when heated too fast or at too high a temperature. Overcooked milk will form a calcium caseinate scum on the top, the whey layer on the bottom, and have an off-flavor. Heating milk in a compartment steamer for a controlled amount of time avoids these unfavorable results.

CURDLING. If milk is held above 185 F for more than a few minutes the milk will begin to curdle.

The curdling is caused by a separation of the casein proteins and the whey. Once milk reaches the desired temperature it should be removed from the heat and used.

Milk also may curdle when it comes in contact with acidic foods, such as tomatoes, citrus fruits, and vinegar. In some recipes this curdling is desired to give the food the distinctive sharp flavor. This combination of milk with acid may also be used to add the acid needed to react with baking soda in baked products. The combination of milk and acid causes a chemical reaction that changes the flavor of both the milk and the acid and results in a sour milk flavor. If the same amount of acid food was

MILK AND CHEESE 95

not combined with the milk before being added to the recipe a different flavor would be noticed. For example if 2 tablespoons of vinegar are combined with 1 cup of milk before being added to a corn bread recipe, the bread will have a slight sourdough- or buttermilk-type flavor. If the vinegar was added to the ingredients separately from the milk, a vinegar flavor would result.

When curdling is not desired, a very small amount of an alkaline ingredient can be added to the acid before the milk. Adding baking soda to tomato before adding milk limits the acid reaction, for example. While this technique reduces curdling, it may cause a bitter flavor and lower the ascorbic acid content. Using a white sauce instead of plain milk can help prevent curdling and is recommended instead of adding an alkaline.

Vegetables that usually curdle milk are tomatoes, asparagus, carrots, green and wax beans, and peas. Vegetables that usually do not curdle milk are cabbage, cauliflower, and broccoli.

COOKING WITH DRY MILK POWDER. Dry milk powder and water may be substituted for liquid in many recipes. When milk is a major ingredient. as in white sauce and cream of mushroom soup, the powdered milk flavor may be detected. The characteristic flavor of powdered milk can be limited by thoroughly mixing the milk powder with water before adding other ingredients. This is necessary only when milk is the main ingredient in a recipe. If milk is used in smaller quantities, the powder should be added with other dry ingredients and the water with the other liquid ingredients.

Two stages happen when dry milk powder is mixed with water. First the powder is suspended in the water. This is much like the slurry made from cornstarch and cold water. If allowed to stand, much of the milk powder will settle to the bottom. In the second stage, the milk powder is dissolved in the water. After it is dissolved it will not settle to the bottom. To obtain the best milk flavor and the usual results of using fresh milk in cooking, the powder must be dissolved in the water.

To thoroughly dissolve the milk powder, the milk must be stirred with a wire whip. In small quantities up to 2 gallons, this can be done by hand. In larger quantities using a mixer is better. The stirring should be slow and constant, bringing the whip in a figure eight pattern from the bottom to the top of the mixture. Fast whipping should be avoided because it will cause the milk to foam. Foamed milk is hard to measure and add to the recipe. In quantity food production, the amount of powdered milk needed for one day can be prepared all at once and refrigerated until needed. It should not be held more that 24 hours before being used. See the recipe for Reconstituting Nonfat Dry Milk and Table 6.2, which gives measurements for making from 1 quart to 6 gallons of fluid skim milk.

Table 6.2. Measures for mixing nonfat dry milk

Fluid Skim Milk	Nonfat Dry Milk			Water
	Weight	Measure		
		Noninstant	Instant	
1 qt	3½ oz	¾ c	1⅓ c	3¾ c
2 qt	7 oz	1½ c	2⅔ c	1 qt 3½ c
3 qt	10½ oz	2¼ c	1 qt	2¾ qt
1 gal	14 oz	3 c	1 qt 1⅓ c	3¾ qt
2 gal	1 lb 12 oz	1½ qt	2 qt 2⅔ c	1 gal 3½ qt
3 gal	2 lb 10 oz	2¼ qt	1 gal	2 gal 3¼ qt
4 gal	3 lb 8 oz	3 qt	1 gal 1¼ qt	3¾ gal
5 gal	4 lb 6 oz	3¾ qt	1 gal 2¾ qt	4¾ gal
6 gal	5 lb 4 oz	1 gal 2 c	2 gal	5 gal 2½ qt

Source: Adapted from USDA (1971).
Note: To determine the volume of instant nonfat dry milk to use in a recipe, multiply the volume of noninstant nonfat dry milk given in the recipe by the factor 1.8.
Nonfat dry milk donated to schools is intended for use in cooking and is fortified with vitamins A and D. One cup (reconstituted) provides 500 international units of vitamin A and 100 international units of vitamin D.

Reconstituting Nonfat Dry Milk

Yield: 1 gal reconstituted

Ingredients	Weight	Measure	Procedure
Fluid Skim Milk Noninstant nonfat dry milk .... or Instant nonfat dry milk .... Water, room temperature	14 oz or 14 oz	3 c or 1 qt 1⅓ c 3¾ qt	1. Sprinkle noninstant dry milk on top of water and beat with mixer, rotary beater, or wire whip until smooth. 2. Mix instant dry milk and water in a gallon jar with a tight lid or a large pitcher. Shake or stir to mix. 3. If not used immediately, cover and refrigerate. *Note:* To prepare sour milk, use 1 c vinegar in place of 1 c of the water in fluid skim milk recipe.
Buttermilk Noninstant nonfat dry milk .... or Instant nonfat dry milk .... Lukewarm water .... Buttermilk, commercial	14 oz or 14 oz	3 c or 1 qt 1⅓ c 3 qt 2 c	1. Reconstitute nonfat dry milk with lukewarm water. Stir in buttermilk. 2. Cover. Let stand at room temperature 8 hr. Stir until smooth. 3. Cover and refrigerate until used.

WHITE SAUCE AND SOUP. Milk is the main ingredient in white sauces and cream soups. Basic proportions for some of these sauces and soups are given on pages 96 through 100.

White Sauce, Using Fresh Milk

Sauce	Use	Fat	Flour	Salt	Milk
Yield: 1 gal Thin	Soup Soup from starchy vegetables	¾–1 c *or* 6–8 oz ½–1 c *or* 4–8 oz	1½–1¾ c *or* 6–7 oz 1 c *or* 4 oz	1½ TBSP	1 gal
Medium	Creamed dishes	1–2 c *or* 8 oz to 1 lb	2–2½ c *or* 8–10 oz	1½ TBSP	1 gal
Thick	Croquettes	1½–2 c *or* 12 oz to 1 lb	3–4 c *or* 12 oz to 1 lb	1½ TBSP	1 gal
Yield: 1 c Thin	Soup Soup from starchy vegetables	1 TBSP 1 TBSP	1 TBSP ½ TBSP	¼ tsp	1 c
Medium	Creamed dishes	2 TBSP	2 TBSP	¼ tsp	1 c
Thick	Croquettes	3 TBSP	3 TBSP	¼ tsp	1 c

White Sauce, Medium, Using Nonfat Dry Milk

Ingredients	Yield					
	1 qt	2 qt	3 qt	1 gal	1¼ gal	1½ gal
Margarine or butter	3 oz	6 oz	9 oz	12 oz	15 oz	1 lb 2 oz
Flour, all-purpose	1½–2 oz	3–4 oz	4½–6 oz	6–8 oz	7½–10 oz	9–12 oz
Salt	1⅛ tsp	2¼ tsp	1 TBSP 1⅜ tsp	1 TBSP 1½ tsp	1 TBSP 2½ tsp	2 TBSP ¾ tsp
Pepper, white	¼ tsp	½ tsp	¾ tsp	1 tsp	1¼ tsp	1½ tsp
Water, cool	¾ c	1 qt 3½ c	2 qt 3¼ c	3¾ qt	1 gal 2¾ c	1¼ gal 2½ c
Milk, nonfat dry	4 oz	8 oz	12 oz	1 lb	1 lb 4 oz	1 lb 8 oz

Procedure

1. Make roux:
   a. Melt margarine or butter in saucepan.
   b. Add flour, salt, and pepper; blend.
   c. Simmer for 3–5 min, depending upon volume.
2. Measure water into steam-jacketed kettle.
3. Add nonfat dry milk. Whip until smooth.
4. Heat milk to scalding (185 F).
5. Add roux to hot milk while stirring.
6. With steam turned low, cook, stirring occasionally, until thickened.

Note: If sauce is to be served immediately, use the larger quantity of flour; however, if sauce is to be held, use the smaller amount, since sauce thickens upon standing.

Veloute Sauce

Yield: 100 portions
Portion size: 1½ oz or 2-oz ladle (scant)

Amount	Ingredients	Amount	Procedure
....	Margarine or butter	1 lb 4 oz	1. Make roux:
....	Flour, all-purpose	10 oz	a. Melt margarine or butter in saucepan.
....	Salt	1 TBSP	b. Add flour, salt, and pepper; blend.
....	Pepper, white	¼ tsp	c. Simmer for approx 3 min.
....	Stock, chicken	1 gal 1 c	2. Measure chicken stock into steam-jacketed kettle. 3. Heat to simmering. 4. Add roux to hot stock while stirring. 5. Cook stirring occasionally, until thickened. 6. Serve hot sauce over chicken or turkey turnovers.

Mushroom Sauce (White Sauce Base)				
Yield: 1 gal				
Amount	Ingredients		Amount	Procedure
	Mushrooms, fresh .... or Mushrooms, stems and pieces, 1-lb can		1 lb 4 oz or 1 can	1. Prepare fresh mushrooms by cleaning, trimming, and slicing. or Drain canned mushrooms.
	Margarine or butter .... Salt .... Onions, dehydrated		4 oz ½ tsp ½ oz	2. Melt margarine or butter. 3. Add salt, onions that have been rehydrated, and mushrooms. 4. Sauté lightly.
	White Sauce Margarine or butter .... Flour, all-purpose .... Salt .... Pepper, white .... Water, cool .... Milk, nonfat dry		10 oz 5–6 oz 1 TBSP 1 tsp ¾ tsp 3¼ qt 14 oz	5. Make White Sauce. See White Sauce recipe for procedure. *Note:* If sauce is to be served immediately, use the larger quantity of flour; if sauce is to be held, use the smaller amount, since sauce thickens upon standing.
				6. Add sautéed mushrooms and onions to sauce. 7. Allow to stand to develop flavor.

Source: *Standardized Quantity Recipe File* (1971).

Today, many quantity production facilities use canned or frozen soups instead of making them from recipes. It is possible to purchase concentrated soups that equal those that could be made. Lower-quality commercially made soups also are sold. The quality and cost in ingredients and labor should be considered in selecting the best soups for the facility.

WHIPPING MILK PRODUCTS. The protein and water in milk will form a thin film that will trap air in small bubbles. A foam made from fresh milk is not stable. As soon as whipping stops, the foam begins to lose its volume and return to liquid milk. Evaporated milk can be whipped into a more stable foam that has a larger volume. To produce the most stable foam possible from evaporated milk, the milk should be 30–32 F. The addition of an acid, like lemon juice or cream of tartar, at the end of whipping also increases the foam stability. Evaporated milk foams are not stable enough to replace whipped cream in quantity recipes. The addition of gelatin further increases the stability and volume, making the foam useful in quantity food production.

CHEESE

Cheese has been produced for more than 4,000 years. It has been a way to preserve fresh milk for later use. It is made by coagulating milk, cream, skim milk, buttermilk, or any other type of milk. The process of cheese-making concentrates the protein and fat and eliminates the carbohydrate.

Cream of Mushroom Soup

Yield: 100 portions *or* 5 gal
Portion size: 6-oz ladle

Amount	Ingredients	Amount	Procedure
	Mushrooms, fresh	5 lb	1. Wash mushrooms thoroughly; slice ⅛" thick.
	Onions, fresh, finely chopped	1 TBSP	2. Melt butter in steam-jacketed kettle.
	Butter or margarine	1 lb 14 oz	3. Add mushrooms and onions; sauté until tender.
			4. Drain. Reserve fat for roux.
	Flour, all-purpose	15 oz	5. Add flour, salt, and pepper to the reserved butter. Mix thoroughly and cook 3–4 min.
	Salt	1¾ oz	
	Pepper, white	1¼ tsp	
	Chicken stock	3½ gal	6. Measure chicken stock into steam-jacketed kettle. Heat to simmering.
	Note: If chicken broth lacks flavor, a chicken base may be used if desired.		7. Add roux slowly to stock while stirring.
			8. Cook, with heat low, until thickened and smooth, stirring frequently.
			9. Add sautéed mushrooms and onions.
	Milk, whole	2½ qt	10. Combine milk with half and half; add to steam-jacketed kettle; mix.
	Half and half	2½ qt	11. Allow to stand to develop flavor (15 min).
			12. Heat to serving temperature (170 F).

Purchasing Guide

Food as Purchased	For 96 Portions	For ____ Portions
Mushrooms, fresh	4 lb 3 oz	
Onions, mature	1 oz	

Source: *Standardized Quantity Recipe File* (1971).

Cheese has a higher protein and fat content, and a lower carbohydrate content, than the same weight of milk.

The curd and whey are separated in the cheese-making process. The curd is allowed to ripen by natural enzyme action, or microorganisms or animal substances are added to the curd. The casein proteins serve as an emulsifier in cheese, preventing the separation of the protein and fat.

Milk is curdled for cheese production by using acid, rennin or a controlled bacteria culture. Rennin and rennet are animal substances and therefore might be avoided by vegetarian clients.

Types of Cheese Products

NATURAL CHEESE. Natural cheese is any cheese made by clotting milk to form a curd and then draining off the whey. Different types are produced by varying the curd concentration and by aging with or without the addition of microorganisms or other flavoring ingredients.

Vegetable Chowder			
	Yield: 100 portions *or* 5 gal Portion size: 6-oz ladle			
Amount	Ingredients	Amount	Procedure
....	Margarine or butter .... Onions, dehydrated .... Parsley, fresh	1 lb 4 oz 5 oz 1 oz	1. Melt margarine in stream-jacketed kettle; add rehydrated onions and finely chopped parsley. Sauté until onions are golden.
....	Flour, all-purpose .... Salt .... Pepper, white	10 oz 2½ oz 2½ tsp	2. Add flour, salt, and pepper. 3. Mix thoroughly and cook over medium heat for 3–4 min.
....	Chicken base .... Milk, 2%	3 oz 2 gal 2 qt	4. Lower heat. Add soup base and milk gradually while blending with a wire whip. 5. Heat slowly to 185 F.
....	Potatoes, russet .... Water, hot .... Zucchini, fresh .... Corn, whole kernel, drained ..	4 lb 8 oz 3 qt 8 lb 2 oz 4 lb 2 oz	6. Peel potatoes and cut into ½" cubes, using power dicer. 7. Cover potatoes with hot water; steam for 15 min; drain. 8. Cut zucchini into 1¼×¼" julienne strips. 9. Add drained potatoes, zucchini, and drained corn to soup. 10. Heat to 185 F; serve immediately.
....	Variation Broccoli, fresh	8 lb 2 oz	11. Prepare broccoli by removing florets and cutting into ½" pieces. Cut tender portion of the stem into 1¼×¼" julienne pieces. Add to soup with drained corn and potatoes.
	**Purchasing Guide**			
	Food as Purchased	For 96 Portions	For ____ Portions
	Potatoes, russet, approx	6 lb 12 oz	
	Zucchini, fresh, approx .... *or* Broccoli, fresh, approx	10 lb to 10 lb 8 oz 10 lb to 10 lb 8 oz	
	Corn, whole kernel, golden, vacuum-packed	1 no. 10 can	

Source: Adapted from *Standardized Quantity Recipe File* (1971).

Cheese is classified on the basis of moisture content and the treatment received after the curd is made. The cheese classifications based on moisture level are listed in Table 6.3.

PROCESS CHEESE. *Process cheese* is made by heating natural cheese and adding an emulsifier. Heating during the processing kills microorganisms, which lengthens the storage life, compared with that of natural cheese. The emulsifier makes an important difference when cooking with the cheese. When heated, the fat in natural cheese separates from the protein, making the cheese look curdled, and the protein part of the cheese becomes stringy. In process cheese, the emulsifier prevents this.

PROCESS CHEESE FOOD. *Process cheese food* is similar to process cheese but has water and milk solids added. The additional moisture makes it a very soft food and also lowers the cost of making it. Cheese food is used when a very soft and easily melted cheese is needed, as it melts into a semisoft mass. It can even be used as a very heavy cheese sauce when melted.

PROCESS CHEESE SPREAD. *Process cheese spread* is a process cheese food with even more moisture and stabilizers added. It is spreadable when stored above 45 F. It may also be made from natural cheese with added emulsifiers and water.

Process cheese spread is used when a cheese product has to be soft even when refrigerated. The extra moisture added to this product causes it to stay soft enough to spread when cold. This moisture also causes the flavor to be weaker and milder than other processed cheeses. To make up for the weakened flavor, artificial flavors are sometimes added.

CHEESE PRODUCT. *Cheese product* is a product with a similar flavor, texture, and cooking quality as process cheese. It is made from milk solids and fats plus cheese or process cheese.

Cheese product comes in a wide variety of forms. It can be made to imitate process cheese, process cheese food, or process cheese spread. The imitation products may contain some natural cheese or be made from milk solids, nonmilk fats, and artificial flavors and colors. The quality and nutritional value of imitation cheese varies widely from one brand to another.

COLD-PACK CHEESE. *Cold-pack cheese* is a cheese product made by adding an emulsifier to a mixture of natural cheeses. Cold-pack cheese, also called *club cheese,* differs from processed cheese in that it is not heated during processing. The added emulsifier enhances the spreadability of cold-pack cheese, making it useful for snacks and sandwiches. The texture of this cheese is slightly crumbly and rough, compared with the smooth plasticlike texture of processed cheese.

Table 6.3. Moisture content and description of natural cheeses

Cheese Group	Moisture Content, %	Types of Cheese	Months Aged	Description
Hard	30–40	Cheddar	1–12	White or yellow, crumbly
		Parmesan	> 12	Very hard, used grated
		Swiss	3–10	White smooth with holes
		Emmentaler	3–10	Type of Swiss with small holes
Semisoft	50–75	Muenster	0	Very soft
		Mozzarella	0	Very soft, mild
		Blue/Bleu	2–6	Firm, veined with blue mold
		Camembert	1–2	Very soft, mold-coated rind
Soft	80–85	Cottage	0	Curds moistened with cream

Cooking with Cheese

The moisture and fat content of natural cheeses influence how they react when cooked. The high moisture of cream cheese makes it easy to blend with other ingredients when heated. Low-moisture hard cheese, like cheddar, is more difficult to blend. Aging low-moisture cheese improves its ability to be melted and blended when hot. A young cheddar cheese does not melt as easily and becomes stringier and more curdled than does a well-aged cheddar. Aging high-moisture natural cheese does not improve its cooking properties.

Stringiness, toughness, and fat separation occur when any natural cheese is overheated, either because of a high temperature or a long cooking time. Overheating is the primary cause of the oily appearance of casseroles topped with natural cheese. Adding the cheese to a white sauce or to other ingredients helps prevent the separation of fat from being noticed because as it separates, it mixes with the other ingredients rather than moving to the top of the product.

SUMMARY

Milk and milk products are an important part of many menu items served at home and in quantity-foodservice organizations. They contribute the major amount of calcium to the diet and are also significant sources of protein. Milk can be served as a beverage, as the main ingredient in foods like cream soups and sauces, or as a source of moisture in breads and bakery products.

Cooking milk requires controlled temperatures to avoid scorching, curdling, and surface scum. Using a compartment steamer to heat milk avoids many of these problems. The chemical reaction between milk and acid creates a sharp, sour milk flavor that is desirable in some foods. It can also provide the acid needed for leavening by baking soda.

Natural cheese has agreeable flavor and texture qualities when served cold. Cooking with natural cheese may result in a stringy, curdled-looking product unless low temperatures are used. Processed cheeses and cheese foods offer an alternative that results in a smoother cooked product. However, there are flavor and color differences that may not be desirable.

LEARNING ACTIVITIES

Activity 1: Heating Milk

1. Pour 3 cups of cold whole milk into a small saucepan.

    A. Taste the cold milk and record your impression on the chart.

    B. Heat the milk to scalding, 185 F, and taste it again. Record the appearance and flavor on the chart.

    C. Continue heating the milk until it reaches the boiling point, 212 F, taste it, and complete the chart.

2. What effects do high cooking temperatures and long cooking times have on milk?

MILK AND CHEESE 103

Activity 1, Step 1

	Appearance	Flavor
Cold whole milk		
Milk after scalding (185 F)		
Milk after boiling (212 F)		

3. How could you use each of the following items to prevent a milk-based soup from curdling?

   Cooking temperature? _____

   Cooking time? _____

   Number of batches? _____

   Holding technique? _____

Activity 2: Comparing the Effects of Heat and Cooking Time on Four Types of Cheese

1. Have ready a slice of each of these cheeses—natural cheddar cheese, process cheese, process cheese food, and process cheese spread—and 4 slices of white bread.

   A. Toast the four slices of white bread.

   B. Place one slice of cheese on each slice of toast.

   C. Cut each slice of the toast-cheese combination into four sections.

   D. Cook one quarter-section of each type of the toast-cheese combination according to the heating temperatures and cooking times listed in the chart.

2. Evaluate the results and complete the chart.

3. Which of the four cheeses heated to the soft stage did your prefer? Why? _____

   _____

4. Which cheese was stringiest? At what cooking temperature and time was this most obvious?

   _____

Activity 2, Step 2

Cheese Type and Quality Evaluation	Cooking Temperature and Time			
	300 F		500 F	
	Just until soft	Cook 5 more min	Just until soft	Cook 5 more min

Process
   Flavor
   Appearance
   Stringiness
   Fat separation				

Natural
   Flavor
   Appearance
   Stringiness
   Fat separation				

Process cheese food
   Flavor
   Appearance
   Stringiness
   Fat separation				

Process cheese spread
   Flavor
   Appearance
   Stringiness
   Fat separation				

5. Which cheese had the most fat separation? At what cooking temperature and time did this occur?

6. Which of the four cheeses would you select to use as a melted topping for a tuna noodle casserole? Why?

REVIEW QUESTIONS

Multiple Choice

1. Whole milk contains at least what percent butterfat?

    A. 1.25%
    B. 2.25%
    C. 3.25%
    D. 4.25%

2. Lactose intolerance is the result of

    A. Too much lactose in the diet
    B. A deficiency of the lactase enzyme
    C. The natural processing of cheese
    D. Too much lactase in the diet

3. Milk products are usually stored in cardboard cartons and heavily frosted jugs because

    A. The insulating effect of the container keeps milk products cooler.
    B. Iron in the milk is destroyed by sunlight.
    C. Riboflavin in the milk is protected from the sunlight.
    D. Carotene is insulated from the sunlight.

4. Pasteurization is used to treat milk in order to

    A. Stabilize the vitamins in the milk
    B. Prevent flavor changes in the milk
    C. Kill microorganisms that cause spoilage
    D. Remove some of the water

5. Which of the following causes milk to curdle?

    A. Acids
    B. Heat
    C. Overcooking
    D. All of the above

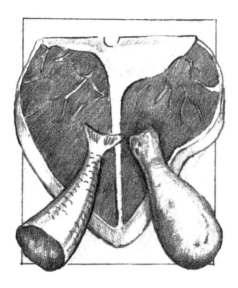

7. MEATS, POULTRY, FISH, AND ENTREES

MEATS, POULTRY, AND FISH

Nutrient Content

Meats, poultry, and fish are major sources of protein. Meat also contains a large amount of iron and several B vitamins; poultry is a good source of the B vitamin *niacin*. Some legumes (primarily beans), corn, rice, cheese, milk, and eggs also provide protein and are the major sources of protein in vegetarian diets.

The entree for a meal is usually a meat or poultry product. The remainder of the meal is planned to complement this main dish. It is very important to offer an acceptable and popular entree. Standardized recipes should be used to ensure portion control and to keep cooking losses to a minimum. The main dish is usually the most expensive item served for each meal.

Composition

Meats, including poultry and fish, are composed of muscle, connective tissue, fat, and bone. The muscle portion of most meat, poultry, and fish is composed of 75% water and 20% protein. The remaining 5% is a combination of carbohydrate, fat, vitamins, and minerals. The ability of these items to hold water and contain fat affects their juiciness. The water content varies according to the variety of product and the type of muscle.

Collagen, an important protein found in meat and poultry, forms the basic structure of connective tissue. Collagen, often referred to as the white connective tissue, is the structure that is broken down by the application of heat, especially moist heat. The greater the breakdown, the more tender the final product.

Elastin, another connective protein, is found in concentrated deposits appearing as a yellow, rubbery mass. It often is referred to as *gristle* and is changed very little by cooking.

Tenderness after cooking has a direct relationship to the amount and quality of connective tissue in the product. For example, the loin strip near the backbone in beef cattle is used as a supportive muscle and has little body movement, resulting in muscle cells that contain very little connective tissue. Muscles in the front part of the leg and shoulder, however, have more connective tissue and are less tender.

Fat, which is found between and within the muscles of all meat, poultry, and some fish, contributes to the meat's juiciness and flavor. The fat content within the lean tissue is critical in the grading of meat. *Marbling,* fat that is dispersed throughout the meat muscle, increases the quality, tenderness, and juiciness in beef but is not desirable in cuts of pork because it gives the cooked pork a greasy taste. Although marbling in beef is considered desirable, it increases the fat, cholesterol, and calories of the meat.

It is important to select a cooking method that will maximize the tenderness of the meat while retaining nutrients. Cooking temperature and time affect the amount of shrinkage and the loss of vitamins and minerals in cooked meat. Protein, when subjected to excess heat, continues to toughen, shrink, and lose moisture until cooking is finished. If the temperature is increased, the protein will continue to contract and shrink until all the moisture is squeezed out, leaving a very dry, tough product. This excessive reduction of liquid from the muscle cells also causes essential fatty acids, minerals, and vitamins to be lost.

The weight loss that occurs during cooking affects the yield percentage. Table 7.1 gives the yield percentages for meat, and the calculation of yield for an 8 pound 8 ounce beef roast. The same procedure is used to calculated yields of pork, poultry, and fish. The original purchase is weighed, and losses before and after cooking are measured and subtracted from the original weight.

$$[(\text{original weight} - \text{all losses}) \div \text{original weight}] \times 100 = \text{yield percent}$$

Table 7.1. Weights and yield percentages for an 8 lb 8 oz beef roast

Weight Categories	Losses %	Losses Weight	Percent of Original Purchase Weight	Resulting Weight
Original weight			100.0%	8 lb 8 oz
Loss in trimming	23.5%	2 lb 0 oz		
Trimmed weight[a]			76.5%	6 lb 8 oz
Loss in cooking[b]	11.8%	1 lb 0 oz		
Cooked weight			64.7%	5 lb 8 oz
Loss after cooking (bone and fat)	17.6%	1 lb 8 oz		
Original weight		8 lb 8 oz	100.0%	
Total losses		4 lb 8 oz	52.9%	
Total servable weight[c]			47.1%	4 lb 0 oz

[a]Trimmed weight = original as-purchased weight minus trim loss.
[b]Loss in cooking = trimmed weight minus cooked weight.
[c]Servable weight = cooked weight minus loss in portioning and slicing.

The yield percentage of 47.1% helps to determine the amount of meat to purchase, taking into consideration the losses that occur both before and after cooking. To determine the amount of meat, poultry, fish, or shellfish to purchase, use the following formula:

$$\text{Amount to purchase} = \frac{\text{Number of portions} \times \text{portion size}}{(\text{Yield percentage} \div 100)}$$

Although proper cooking causes some shrinkage and weight loss in meat, it produces many desirable changes:

- It changes the appearance for better eye appeal.
- It kills any surface microorganisms.
- It softens the tissue.
- It develops flavor.

Meat Products

INSPECTING. Inspection of meat must be done by U.S. Department of Agriculture (USDA) inspectors for interstate sale or by state inspectors for sale within a state. The round inspection stamp bears the number of the processing plant and states: "U.S. Inspected and Passed." This shows the meat carcass was safe for consumption at the time of the inspection. Since the federal inspection standards are higher, quantity foodservices should buy only federally inspected meats.

GRADING. Grading is not required either to ship meat between states or to sell it within the state in which it is produced and processed. Quality grade is determined by two factors, maturity and marbling.

Here are the most current grades used for beef, veal, pork, and lamb:

BEEF	VEAL	PORK	LAMB
Prime	Prime	U.S. #1	Prime
Choice	Choice	U.S. #2	Choice
Select	Good	U.S. #3	Good
Standard	Standard	U.S. #4	Utility
Commercial	Utility	Utility	
Utility			
Canners and cutters			

U.S. Choice, the beef grade most often purchased by quantity foodservices, is the grade most abundantly produced in the United States. However, a lower grade like U.S. Select is often chosen for use in low-fat entrees because it has a lower fat content than Choice, just as Choice has a lower fat content than Prime. The lower grades are generally better choices for low-fat and low-cholesterol diets.

Pork grades are based more on yield than on quality, because there are minimal differences between the grades for flavor and tenderness. Fresh pork that is watery or soft is unacceptable. During recent years the demand for leaner cuts of pork has prompted breeders and producers to market carcasses that contain less fat. Pork can no longer be considered higher in fat content than beef.

Grades used for veal are very similar to those in beef. The higher veal grades have higher levels of fat and thicker flesh than do lower grades. Veal bones are small in relationship to the weight and size of the carcass, but because of the high cost of veal, few foodservices can afford to offer it on a regular basis.

When lamb is offered on the menu, Choice or Good grades are usually selected. Mutton is rarely used, due to its strong flavor and odor.

PURCHASING. All meat products should be purchased by specifications. The National Association of Meat Purveyors publishes *The Meat Buyer's Guide*, which is commonly used by meat suppliers and food buyers to help standardize these specifications. The Institutional Meat Purchase

Specifications (IMPS) numbers, which were developed for use in institutions, are listed in *The Meat Buyer's Guide*. These specifications can aid quantity foodservice managers in ordering the exact meat listed in the recipe.

Meat specifications for purchase should include the following:

- Name of the cut
- USDA grade
- Whether chilled or frozen
- Packaging and/or number of units per container
- Any requirement for boning, rolling, and tying
- The IMPS number

These specifications will ensure that the exact meat product is ordered to meet production needs. Exact specifications should be used no matter what unit size is purchased.

Poultry Products

INSPECTING AND GRADING. Chicken must be inspected by the USDA when it is shipped across state lines. State inspections are required for poultry sold within states. Quantity foodservices should purchase only chicken and other poultry products inspected by the USDA, to ensure that strict sanitary standards were met during processing. If inspected by the USDA, the federal inspection circle mark will appear on the package label.

Plant inspections help prevent contamination by salmonella, a common bacteria often present in raw poultry products, and to ensure that proper sanitation standards are maintained. If poultry contaminated by salmonella is improperly handled and prepared, massive outbreaks in foodborne illnesses can occur.

Poultry may be graded as U.S. Grade A, U.S. Grade B, or U.S. Grade C. U.S. Grade A is most commonly used in quantity foodservice operations. The lower two grades do not have a high quality of meat conformation, a well-developed layer of fat, or skin free from cuts and defects. Lower grades might be appropriate for use in casseroles where the whole chicken section is not being served. Diced chicken or turkey can be purchased frozen for use in salads and casseroles.

Most types of ready-to-cook poultry are available as whole, halved, or quartered forms. All these forms require washing and trimming of excessive fat and skin before it is cooked.

PURCHASING. Poultry specifications should include the following:

- The species of bird, such as turkey, chicken, or duck
- The class, which means age and sex. Classes are as follows:

   Broilers: Young birds, primarily used for broiling, 1¼–2½ pounds, very tender and little flesh
   Fryers: Young birds, 2½–3½ pounds, used for frying, roasting, and broiling
   Roasters: Quick-grown birds, 3½–5 pounds, primarily used for roasting
   Capons: Unsexed males that grow rapidly would fall under this class
   Roosters or hens: Older birds, 3½ pounds or more, used for stewing, soup stock, and some casseroles.

Fish and Seafood Products

INSPECTING AND GRADING. Fish processors are inspected on a voluntary basis by the National Marine and Fisheries Service of the U.S. Department of Commerce. Inspected fish products display a shield stating the grade. Fish is graded as Grade A, Grade B, or Grade C. Grades A and B are the most common. Grade B has more variation in size and more blemishes than Grade A.

Fresh fish should have shiny and unfaded skin, clear eyes, and red gills. The odor should be fresh and not strong. Frozen fish should be packaged in airtight containers and frozen solidly. Fish is often sold frozen in solid blocks that require thawing and portioning. It can be specified to arrive frozen in individual breaded, battered, or unbreaded fillets. These preportioned fillets make it easy for quantity foodservices to standardize portion sizes.

Some fish products, such as frozen unbreaded shrimp or scallops, are packaged in a frozen block of ice. Unless specified as individually quick frozen (IQF), the yield of edible product from the solidly frozen block may be less than 25% of the total frozen received weight because considerable water is frozen with the seafood. Therefore, the foodservice department purchaser should pay attention to the pack when placing the initial order, especially when making price comparisons between frozen-block and IQF fish products.

Fish are divided into two types, finfish and shellfish. Finfish are considered either lean or fat. Fish that are less than 5% fat are classified as *lean*. All other fish are classified as *fat*. Shellfish, the other major type, are classified as *crustaceans* (with jointed, horny outer covering) or *mollusks* (with shells). Here are examples of the four major types of fish:

Lean Fish	Fat Fish	Crustaceans	Mollusks
Cod	Tuna	Crabs	Clams
Halibut	Salmon	Lobsters	Oysters
Flounder	Mackerel	Shrimp	Scallops
Perch			

All fish are highly perishable and deteriorate rapidly. The fresher the fish, the better the final cooked product.

Storage

FRESH BEEF AND PORK. Meat should be refrigerated at 30–36 F, in a high relative humidity of 80–90%. Fresh meat should not be stored for periods longer than 3–4 days. When refrigerated, it should be completely covered, yet loosely wrapped. The more surface area of meat exposed to air, the faster it deteriorates, but *loose* wrapping prevents the growth of bacteria that thrive in the absence of air.

Meat should be stored away from other foods, such as vegetables and fruits, to prevent cross-contamination. Hams and cured meats should be wrapped or covered to prevent their odors from being absorbed by other foods in the refrigerator.

Sliced, ground, and cubed meats should be used within 24 hours.

FROZEN BEEF AND PORK. Frozen meat should be tightly wrapped in airtight containers, freezer paper or plastic, or aluminum foil. This prevents the exposure to air that would cause dehydration and freezer burn. It also prevents the absorption of flavors from other foods in the freezer. Such tight wrapping allows beef to be stored at 0 F or below for up to 12 months. The fat of pork products may become rancid and/or develop off-flavors if stored longer than 6 months. Frozen meat that has been thawed should be used at once and never refrozen. Partially thawed meat that still contains ice crystals can safely be refrozen, but the refreezing will cause excessive loss of juices and off-flavors, resulting in a much lower-quality product.

FROZEN VEAL AND LAMB. Unlike beef, which can be stored frozen for 12 months, frozen veal and lamb are limited to 6–9 months storage. If stored longer, the meats may become rancid and/or develop off-flavors.

POULTRY. Like meats, poultry should be stored under refrigeration as soon as it is received. The threat of salmonella contamination makes immediate refrigeration of fresh poultry critical. Most poultry products are purchased frozen. Ideally, they should be thawed just prior to preparation.

When poultry is thawed or stored in the refrigerator, it should be placed in a location where its juices cannot contaminate other foods. Raw poultry products should never be placed in areas of the refrigerator where leftovers, dairy items, eggs, fruits, or vegetables are stored.

Thawing large poultry products such as turkey at room temperature is very dangerous. The outside of the poultry can reach room temperature while the inside is still frozen, which allows microorganisms to multiply on the outside.

FISH. Fish products must be refrigerated as soon as they are received. Fresh fish should be prepared the same day it is received and discarded after 72 hours. Frozen fish products are the best quality if used within 3 months, but frozen fish can be stored safely for up to 6 months.

Meat Preparation and Cooking

PREPARATION AND COOKING METHODS. The decision of whether to use moist- or dry-heat cooking methods depends on the tenderness of the meat. To cook less tender cuts, enzymes like *papain* (found in papaya) or *bromelin* (produced from fresh pineapple) can be used to help tenderize the meat by breaking down the connective tissue. Mechanical tenderization, including grinding, cutting, cubing, scoring, and pounding, can also break down the connective tissue. When untenderized meats with large amounts of connective tissue are being prepared, it is best to use moist cooking methods like braising or stewing.

Tender cuts of meat, such as beef tenderloin, can be prepared by roasting, broiling, panbroiling, pan-frying, or deep-fat-frying. However, even these tender cuts of meat can become tough if they are cooked too long. The more the muscle is heated, the more the protein draws up and becomes tighter, making it tough and difficult to eat.

DRY-COOKING METHODS FOR MEATS. The most common methods for cooking meats are

- Baking or roasting
- Broiling
- Pan- or oven-frying
- Deep-fat-frying
- Barbecuing

Baking or Roasting. Roasting is a common method of dry-heat cookery. Meats can be roasted in conventional, convection-air, and deck ovens. The meat is placed on a rack in the oven with the fat side up to allow it to self-baste. No moisture is added to the uncovered pan. Heat from a conventional oven or the hot moving air of the convection oven produces surface browning, so there is no need to sear a roast before placing it in a roasting pan. Searing only increases total cooking losses, rather than preventing the loss of juices.

The most commonly selected roasting temperatures are 300–325 F. This temperature range produces the tenderest product with the least amount of shrinkage. The standardized recipe should be checked for exact temperature settings.

Broiling. Broiling is another form of dry-heat cooking. The meat is placed directly under or over radiant heat from gas flames, charcoal briquettes, or electric heating elements. This method of cooking is very rapid. To keep meat from becoming dry and tough, it should be turned only once during broiling. Seasoning with salt should be done after the cooking is completed because the salt draws moisture out of the meat and delays browning. Frozen meat can be broiled but should be placed farther away from the heat to prevent the outside from burning before the inside is done. Meat thicker than 1½″ should be thawed prior to broiling.

Pan- or Oven-Frying. In frying, heat is conducted from the surface of the braising pan, grill, or skillet to the surface of the meat. The cooking surface should be oiled before lean cuts of meat are placed on it. Medium cooking temperatures will prevent overbrowning and crusting of meat surfaces. Many recipes, like Swiss steak, combine browning with moist heat. Meat can also be oven-fried. In oven-frying, the meat is placed on a greased sheet pan. The top surface is brushed with fat, and it is cooked at a high temperature.

Deep-Fat-Frying. Meats should be coated or breaded before deep-fat-frying to preserve moisture and add texture and flavor. Careful control of the temperature in deep-fat-frying is very critical. If the fat is too hot, the food will be overcooked on the outside before the center of the meat is done. If the fat is too cool, the longer cooking time will allow increased absorption of fat, leaving a greasy end product. Common frying temperatures are 350–375 F. When large quantities are fried, the fryer must be allowed to return to the proper frying temperature before cooking the next batch. After frying, the product should be placed on a wire rack for draining.

MOIST-HEAT COOKING METHODS FOR MEATS

Braising. Braising is primarily used for less-tender cuts of meat. The terms *fricasseeing, pot roasting,* and *Swissing* also mean braising. Often the meat is dredged in a small amount of flour to increase browning when fried in a small amount of fat. After browning, these meats are cooked slowly in a small amount of liquid. Steam-jacketed kettles and tilting braisers are used in quantity foodservices for this cooking method.

Stewing. Stewing is used to cook the toughest cuts of meat. It is similar to braising except that the meat is completely covered in liquid. To make the product more tender, the meat is simmered, rather than boiled, after the liquid has been added to the recipe. Vegetables can be cooked in the stew or separately and added as a garnish for the meat.

The steps in the stewing cooking method are as follows:

1. Cut the meat into uniform pieces, 1- to 2-inch cubes.
2. Brown the meat at 350 F.
3. Add seasonings: salt, pepper, and spices.
4. Cover and simmer at 300 F until tender.
5. Check tenderness with a fork.

Although less-tender cuts are sometimes cooked in pressure steamers or tightly wrapped foil, shrinkage and drip loss actually may be greater than when using other moist-heat cooking methods. In addition, the flavor and color of steamed meat is not highly desirable.

The final cooked weight is greatly affected by the amount of shrinkage that occurs during cooking. Factors influencing the amount of shrinkage in meat are

- The cut of meat
- The method of cooking
- The cooking temperature
- The length of cooking period
- The degree of doneness

COOKING TIMES FOR MEATS. Tenderness in meat is affected by the length of the cooking time. Factors affecting length of cooking time are

- The cooking temperature
- The oven load
- The degree of doneness
- The tenderness of the cut of meat
- The temperature of the meat prior to cooking
- The size of the meat

CHECKING FOR DONENESS. A meat thermometer should be used to check meat for doneness. The muscle is pierced at its thickest point without letting the thermometer touch the bone. Meat should not be tested for doneness by piercing with a fork or cutting with a knife because this releases the juices from the meat and causes it to become dry.

The final temperature for beef and lamb depends on personal preference. Many people prefer lamb cooked medium rather well done. When roasting meats of various sizes, the internal temperature of the smallest item should be checked first, to prevent it from overcooking. The internal temperature of meats such as roasts will continue to rise even after being removed from the oven. Here are guidelines to determine the three major stages of doneness.

Doneness	Internal Temperature	Description
Rare	145 F	Flesh very red, outside brown and plump; juices red but not bloody
Medium	150 F	Interior of flesh rose and juices pink, but less juice than rare; exterior well browned with flesh that resists pressure; less plump than rare
Well	160 F	Flesh completely cooked with little or no juice; exterior hard with a shrunken look; surface dry and dark brown

Due to improved pork production methods, trichinosis, an infection caused by parasites occasionally found in raw and undercooked pork, is not the threat that it once was. The Foodservice Department of the National Pork Producers' Council recommends a final cooking temperature of 145–160 F for all pork. The USDA's Food Safety and Inspection Service suggests the end point for cooking pork be 160 F.

Poultry Preparation and Cooking

PREPARATION. Poultry products require some cleaning and preparation immediately before cooking.

Organs, such as lungs on chicken breasts, are removed, as are pinfeathers, extra fat, and loose pieces of skin. The joint is cut between the leg and thigh to allow these pieces to be bent easily when panning for oven-cooking. Cutting tendons between the wing and breast allows the pieces to lay flat, which speeds cooking.

All raw poultry products should be washed in cold water to reduce bacteria levels and remove small pieces of fat and organs. Drain off the water, and cool immediately. Cover before storage in the refrigerator.

DRY-HEAT COOKING METHODS FOR POULTRY

Roasting. The same cooking methods used for meats are used for poultry. One of the most common methods in the preparation of poultry is roasting. Turkey can be either by roasted whole or cut into sections to allow for different cooking times of white and dark meat.

Most quantity foodservices use either deboned turkey breasts or turkey rolls that contain white meat or a combination of white and dark meat. Usually these turkey breasts and rolls have been precooked and only require reheating in the compartment steamer. Occasionally they are roasted in a conventional oven when pressurized steamers are not available. If they are cooked in the oven, a small amount of heated water should be added to the pan to help prevent them from drying out.

Turkey rolls are used by many quantity foodservices because they are very convenient. They are easy to slice for entrees or cold sandwiches. But they may not be appropriate for low-sodium diets. Law requires manufacturers to provide nutritional information, such as sodium content, when it is requested. In addition to being higher in sodium content, turkey rolls and breasts are much more expensive per pound than whole turkeys or turkey parts. Some clients object to the flavor and texture of processed turkey rolls.

Using foil used to cover the turkey breasts prevents overbrowning and drying. The foil should be removed to allow for browning during the last 30 minutes of cooking. Turkeys are usually done when the inside temperature is 180 F. When checking for doneness, the thermometer should be inserted into the thigh muscle at the thickest part where it is joined to the body carcass. Oven cooking bags are not usually used in quantity production. They add to the cost and equivalent results can be obtained with the foil tent method.

For turkey to be used in casseroles, salads, and sandwiches, some foodservices simmer the poultry in a steam-jacketed kettle. Although this method prevents overbrowning and may reduce cooking losses, the flavor may not be as desirable as that of roasted turkeys. To simmer poultry, place the thawed, cleaned birds in the kettle and cover with water. Simmer until the meat loosens from the thighbone and the internal temperature at the thigh joint is 180 F. The resulting broth is useful for soups, sauces, and casseroles.

Turkeys must be handled carefully to avoid the growth of microorganisms. They should not be roasted overnight at a low temperature (below 200 F). If not served immediately, they should be refrigerated and never allowed to remain at room temperature for more than 1 hour. However, if being carved for immediate service, turkeys should stand for 15 minutes after removal from the oven to allow the meat juices to be reabsorbed.

Quantity foodservice managers should never allow whole turkeys to be stuffed with dressing. Poultry cooks and cools faster without the stuffing. The extended cooking and cooling times needed for stuffed poultry allows microorganisms to multiply and contaminate the final product. It is better to bake the dressing separately in shallow pans that will allow the dressing to be cooked and cooled rapidly.

Broasting and Slow Roasting. Broasting has become a popular method for cooking poultry, especially chicken. The broaster applies steam under pressure while the chicken is being deep-fat-fried. The end product is a very tender and juicy meat. Other equipment, like slow roast and hold units, can be used for poultry by following the manufacturer's cooking instructions.

Frying. Frying is a popular method to prepare young and tender chicken. Frying can be done in the oven on greased sheet pans or in the deep-fat-fryer. The oven-fried method, which more accurately could be called baking, is most practical for preparing large quantities. Usually oven-fried chicken is dipped in a milk or margarine mixture before being lightly coated with bread or cracker crumbs and spices. The dip and coating depends on the recipe.

Deep-Fat-Frying. Deep-fat-frying is used for chicken pieces, processed chicken patties, chicken tenders, and a variety of convenience chicken products. These products are fast and easy to prepare. The temperature of the fat for frying these products should be 350–365 F. Higher temperatures will cause the product to become too brown, and lower temperatures promote grease absorption.

When deep-fat-frying chicken, a breading or batter coating is used to create an appealing, crisp outside and a juicy interior. If the chicken pieces are of various sizes, the largest pieces are placed in the fryer first, followed by the smaller ones. At the correct temperature of fat, all the pieces will be thoroughly cooked and will rise to the surface.

Broiling. Broiling is commonly chosen for the more tender cuts of poultry, like young chicken quarters or chicken breasts. This method is growing in popularity with the increasing public awareness of low-calorie, low-fat cooking methods.

MOIST-HEAT COOKING METHODS FOR POULTRY. Stewing, braising, steaming, and simmering are used for older, less tender poultry products. Whichever method is used, the poultry should be cooked until the meat can be easily removed from the bone with a fork.

Stewing. Stewing is an effective method for cooking poultry for creamed sauces, soups, casseroles, and gravy. With the availability of commercially prepared frozen and canned chicken and turkey, quantity foodservice managers may choose to use the convenience products to save time and to ensure a standardized product.

Fish Preparation and Cooking

Because fish deteriorates very rapidly, it should be kept refrigerated until it is prepared. Frozen breaded fish should not be allowed to thaw. Instead, it should be placed directly in the oven or the deep-fat-fryer. Lean fish are best when poached or steamed. Fat fish are best when broiled, pan-fried, or baked. Shellfish should be broiled, sautéed, poached, steamed, or simmered for a short period of time to avoid toughening the protein. Oven-baking, oven-frying, and deep-fat-frying are common methods that require the least amount of time and labor. Pan-frying and broiling are more labor intensive.

Here are considerations in selecting cooking methods for fish:

- Lean fish should be cooked at low temperatures.
- Fat fish should be broiled or fried.

When baking fish in the oven, it is best to place unbreaded fish fillets on a well-oiled sheet pan. In oven-frying, breaded fish fillets are also placed on a well-oiled sheet pan but with a thin layer of melted shortening poured over the top. Fish products are done when the flesh is white and flakes when tested with a fork.

LEFTOVERS

Quantity foodservices should be managed so the amount of leftovers is minimized. Even with accurate forecasting, however, all operations have some leftovers. Because the quality of prepared foods deteriorates very rapidly, leftovers must either be stored under refrigeration or frozen for later use, or be disposed of properly.

Protein-containing foods should be refrigerated or frozen to inhibit growth of microorganisms. Leftovers should be inventoried by the foodservice manager or production supervisor to ensure that they are rotated and used within 48 hours.

Leftovers should be tightly covered, labeled, dated, and stored away from all raw and uncooked foods. If they are frozen, they should be in heavy foil, freezer wrap, or approved plastic or metal containers. They should never be stored in containers in which the food was originally received. Containers covered only with aluminum foil or plastic film should not be stacked, but instead put on a rack with only one layer of pans per shelf. If necessary, 12×20" pans can be stacked in alternating directions so the pan edge, not the cover, is supporting the weight of the upper pan.

When leftovers are placed in the freezer, space should be left between and beside pans for air circulation until all pans are solidly frozen. Stacking unfrozen foods too close together will cause them to take longer to freeze, and placing too many room temperature and warmer items in a freezer or refrigerator may raise the temperature of the freezer for several hours. This may cause foods to partially thaw, spoil, or become contaminated. Microorganisms can grow at this temperature, contaminating the thawed food. Foods on lower shelves also may be contaminated by microorganism-laden moisture dripping down.

Refrigerators and freezers should be organized so leftovers can be retrieved easily. One example is placing all leftover beef products on one rack and all leftover pork products on another. A diagram of the leftover storage areas placed on the freezer door will help to keep order.

Leftovers should be reheated to 165–170 F before they are served again. Once they are reheated, any that are not eaten should be disposed of properly and never be refrozen or refrigerated. Leftover casseroles, gravies, dressings, and any deep-dish foods should be placed in shallow containers to allow for rapid cooling. Foods refrigerated in deep containers cool slowly and allow microorganisms to grow rapidly in the warm center part of the product.

Leftover meat dishes containing tomato products should not be stored in aluminum foil because the acid from the tomatoes will dissolve the foil. Such products should be covered by a layer of plastic film and then foil or with a tight-fitting stainless steel or plastic lid.

SANDWICHES

Sandwiches are a very popular entree with most clients. These food combinations make it possible to offer a large variety of breads, spreads, and fillings.

Close-grained bread should be used for sandwiches because it is less likely to become soggy from moist fillings, and it does not stale as fast as coarse breads. Sandwich fillings should provide color, flavor, and texture and be nutritious. Fillings can be spreads, meats, cheeses, poultry, or sausages.

When any turkey hams or sausages are used for sandwiches, the fat content should be checked if the type of content or number of calories from fat is important. Many of these products, although assumed to be low-fat by many consumers, are actually higher in fat content than similar beef and pork products.

Sandwich fillings often use salad dressings to hold the ingredients together. Two common examples are tuna and ham salads.

Sandwiches made from egg products along with a salad dressing with a mayonnaise base can provide a medium for microorganism growth. These sandwiches should be refrigerated if not served immediately. If not served within 24 hours, the leftover sandwiches should be disposed of properly in the garbage disposal or sealed and placed in the dumpster.

If sandwiches are to stand for several hours, they should be covered to prevent the bread from drying out and the absorption of off-flavors. They are also much easier to handle if they are wrapped.

Sandwich Production

Sandwich production flow charts should be drawn for efficient use of labor and food. Good planning and work-station organization will facilitate the speed of production. With a well-planned flow chart, less-skilled personnel can be used in the assembly, freeing the more-qualified cooking personnel for other production tasks. Many foodservice managers use a deli-style assembly to allow for a wide variety of sandwich combinations using the many kinds of sliced meats and cheeses.

The sandwich work center should be comfortable for the worker. A 34-inch working height is recommended for short workers. The height should be 36–42 inches for tall workers. Plates, necessary tools, and utensils should be within easy reach of the worker.

Portion control of sandwich filling is very important. Recipes should be closely followed, and appropriate dipper sizes should be used to ensure uniform amounts of filling. For example, if a No. 20 dipper is used (1½ ounces) for a sandwich spread, 100 sandwiches can be made from 5 quarts of filling.

Grilled or fried sandwich patties can be prepared separate from the bread or bun. The final assembly can be completed on the serving line.

Grilled sandwiches, like grilled cheese or Rueben, should be cooked as close as possible to serving time to ensure maximum freshness and quality. Sandwiches that are usually grilled can also be prepared by placing them on sheet pans and baking them in the oven. This is an effective method for producing a large quantity of sandwiches with limited labor.

SUMMARY

A cooking method that will maximize the tenderness of meats, poultry, and fish while retaining nutrients should be chosen. Cooking temperature and time affect the amount of shrinkage and loss of nutrients. Excessively high heat or a long cooking time increases shrinkage and reduces the tenderness of meat. Slow, low-temperature cooking helps to ensure higher quality and yield with a more uniform doneness.

To ensure wholesomeness of meats, poultry, and fish, foodservices should buy only meats that have been inspected. Proper storage conditions will help retain high-quality products.

Meat grades are determined by marbling and the maturity of the animal. Specifications for meats, poultry, and fish must be used to make sure that exact products are ordered to meet the foodservice's needs.

If leftover protein foods are going to be used at a later time, they need to be stored under refrigeration to inhibit growth of microorganisms. A planned procedure for storing and using leftovers should be used.

Sandwiches are a popular entree choice for clients. Many products are available as filling ingredients. Portion control of the filling ensures that uniform amounts are served for the clients.

LEARNING ACTIVITIES

Activity 1: Cooking and Comparing Two Beef Roasts

1. Plan a menu that includes roast beef.

    A. Use two similar roasts of about 10 pounds each and from the same cut of meat.

    B. Follow the directions on the Beef Roast recipe. Roast 1 will be roasted at 300 F and roast 2 at 425 F (use the same recipe, only changing the oven temperature and cooking time).

    C. Insert a meat thermometer into the center, thickest part of each roast before cooking.

    D. Remove each roast from the oven when it reaches an internal temperature of 150 F.

Beef Roast (Round)				
Yield: 100 portions Portion size: 3½ oz			Roasting temperature: 300 F Roasting time: See Procedure	
Amount	Ingredients		Amount	Procedure

…………	Beef, boneless round, choice grade		30–32 lb	1. Place defrosted, uncooked roasts, fat side up, on racks in roasting pans. Avoid crowding roasts.
				2. Insert meat thermometer into each roast so bulb is at center of largest muscle.
				3. Do not add water.
				4. Place uncovered roasting pans in oven at 300 F.
				5. Roast to the desired degree of doneness.
				  Internal Approx
				  <u>Temperature</u> <u>Min per lb</u>
				  Rare …………………… 20–25 min
				  Medium ……………… 28–32 min
				  Well-done …………… 35–38 min

	Weight, cooked		24 lb	6. Remove roasts from pan and allow to stand 15–20 min before slicing, or cool completely if to be sliced very thin for Italian Beef Sandwiches.

Source: Adapted from *Standardized Quantity Recipe File* (1971).

2. Complete the chart.

Activity 1, Step 2

	Roast 1 (300 F)	Roast 2 (425 F)
Weight before cooking		
Edible weight		
Number of 3-oz portions possible		
Appearance		

3. Cut a 3-ounce portion from the center of each roast and then complete the chart.

MEATS, POULTRY, FISH, AND ENTREES 119

Activity 1, Step 3

	Flavor	Appearance	Juiciness
Roast 1 (300 F)			
Roast 2 (425 F)			

4. Cut a portion from the end of each roast, compare each with the center cut from the same roast, and complete the chart.

Activity 1, Step 4

	End Piece (flavor, appearance, juiciness)	Center Piece (flavor, appearance, juiciness)
Roast 1 (300 F)		
Roast 2 (425 F)		

5. Weigh and describe the color of the drippings in each pan and compete the chart.

Activity 1, Step 5

	Weight of Drippings	Color of Drippings
Roast 1 (300 F)		
Roast 2 (425 F)		

6. What caused the difference in the cooked weights of the two roasts? _____

7. Was one roast more tender, juicy, and flavorful than the other? _____

8. Which cooking temperature would be better for cooking roasts in your foodservice operation?
_____

Activity 2: Cooking and Comparing Two Meat Loaves

1. Using the standardized recipe, make two recipes of Meat Loaf from 100% beef with no fillers. In Recipe 1, use 70–80% lean beef. In Recipe 2, use 80–90% lean beef. Standardize all production procedures to ensure that the only difference in the two meat loaf recipes is the leanness of the ground beef.

Meat Loaf			
Yield: 96 portions *or* 2 pans Portion size: 48 per pan			Baking temperature: 325 F Baking time: 1 hr
Amount	Ingredients	Amount	Procedure
....	Pork, lean, ground .... Beef, ground	3 lb 11 lb	1. Place ground meat in 30-qt mixer.
....	Bread crumbs, fine dry .... Milk, nonfat, dry .... Salt .... Pepper, black	1 lb 8 oz 5 oz 3 oz 2 tsp	2. Combine bread crumbs, nonfat dry milk, salt, and pepper in a large bowl. 3. Mix thoroughly.
....	Eggs .... Worcestershire sauce .... Tabasco sauce .... Water	1 lb 2 TBSP ¼ tsp 1¼ qt	4. Beat eggs until well mixed. 5. Add sauces and water to eggs; blend.
....	Onions, dehydrated .... Carrots, shredded or chopped fine	4 oz 1 lb 8 oz	6. Rehydrate onions. 7. Shred or chop carrots.
	Total weight	22 lb	8. Add combined liquids, onions, and carrots to bread crumb mixture; mix very thoroughly. 9. Add crumb-liquid mixture to meats in mixer. 10. Mix with flat beater on low speed for 30 sec. 11. Scrape down sides and mix for an additional 30 sec. 12. Scale mixture into greased pans. *Amt per pan:* 11 lb 13. Bake at 325 F for 1 hr. 14. To serve: Cut crosswise through the meat in each pan to form 4 loaves. Slice each loaf into 12 slices. *Note:* If desired, Piquante Topping or tomato sauce may be added to meat before baking.
Purchasing Guide			
Food as Purchased	For 96 Portions		For ____ Portions
Carrots, topped, approx	2 lb		

Source: *Standardized Quantity Recipe File* (1971).

MEATS, POULTRY, FISH, AND ENTREES

2. Fill in the chart.

Activity 2, Step 2

Quantity Evaluation	Recipe 1, 70–80% lean	Recipe 2, 80–90% lean
Weight prior to baking		
Weight of cooked drippings in pan		
Weight loss during cooking		
Cooked weight of meatloaf (excluding drippings)		
Color and appearance		

3. Cut a 3-ounce serving from the center of each meat loaf and complete the chart.

Activity 2, Step 3

Quality Evaluation	Recipe 1, 70–80% lean	Recipe 2, 80–90% lean
Color and appearance		
Taste and flavor		
Tenderness		
Form (Did it keep its shape?)		

4. What caused any difference in the cooked weights of the two meat loaves?

5. What might have been the reason for any flavor or tenderness differences?

6. Did using 80–90% lean ground beef increase or decrease the overall yield?

7. How does fat contribute to the tenderness of the final cooked product?

Activity 3: Cooking and Comparing Two Recipes for Chicken Newburg

1. Using the Chicken Newburg recipe, make two equal recipes for 33 servings. In Recipe 1, use 3 pounds of frozen ½" diced chicken. Recipe 2, use 3 pounds of fresh ½" diced chicken. Recipe 2 will require at least four 2¼-pound chickens to produce 3 pounds of diced cooked chicken. The average cooked yield of edible meat for chickens is approximately 35% (four 2¼-pound fryers = 9 pounds × 35% yield = 3.15 pounds).

Chicken Newburg			
Yield: 33 portions Portion size: 4 oz (4-oz ladle)			
Amount	**Ingredients**	**Amount**	**Procedure**
....	Chicken, cooked	3 lb	1. Tear or pull chicken into bite-size pieces or, if preferred, cut into ½" cubes. Refrigerate until ready to combine.
....	*Sauce* Margarine or butter .... Flour, all-purpose .... Pepper, white	8 oz 6 oz 1 tsp	2. Melt margarine or butter in saucepan; add flour and pepper; blend and simmer for 3–5 min.
....	Half-and-half .... Chicken broth .... Chicken base	1 qt ⅓ c 1 qt ⅔ c 1 oz *or* 2 tsp	3. Measure half-and-half, chicken broth, and chicken base into steam-jacketed kettle. Stir to blend and heat to 185 F. 4. Add roux while stirring; cook until thickened and flour taste disappears.
....	Sherry, dry, per pan	½ to ¾ c	5. Divide and scale into 12×20×2½" pan. 6. Heat chicken to serving temperature in steamer. *Approx time:* 6–10 min 7. Add 1 pan heated chicken to each pan of hot sauce just prior to service; add sherry. Blend gently but thoroughly. 8. Serve in toast cup, on melba toast, or with spoon bread. Garnish.

Source: *Standardized Quantity Recipe File* (1971).

2. Complete the chart comparing the two products.

Activity 3, Step 2

	Frozen Chicken	Fresh-cooked Chicken
Total labor time		
Cost of raw food		
Cost of labor		
Total cost		
Total number of 4-oz servings		
Total cost per serving		
Flavor		
Aroma		
Texture		

REVIEW QUESTIONS

Multiple Choice

1. The center of a piece of meat cooked medium done is

    A. Very red and 140 F
    B. Rose and 150 F
    C. Light brown and 155 F
    D. Rose and 165 F

2. The preferred cooking method for beef tenderloin is

    A. Simmering
    B. Broiling
    C. Frying
    D. Steaming

3. Which of the following is *not* a good test for doneness in roasted chicken?

   A. Loose joints
   B. Internal temperature
   C. Clear and translucent juices
   D. Tender breast when pierced with a fork

4. A federal inspection stamp on a cut of meat indicates the meat is

   A. Tender
   B. Choice quality
   C. Properly larded
   D. Wholesome

5. Tenderness in cooked meat is determined by

   A. The specific cut
   B. The cooking method
   C. The maturity of the animal
   D. All of the above

6. Which one of the following should be purchased to make oven-fried chicken?

   A. Broiler
   B. Hen
   C. Capon
   D. Roaster

7. Fresh fish should have which appearance?

   A. Soft, loose flesh
   B. A fishy smell
   C. Gray gills
   D. Bright, bulging eyes

8. Shellfish with a jointed, horny outer covering are called

   A. Mollusks
   B. Crustaceans
   C. Flatfish
   D. Fumets

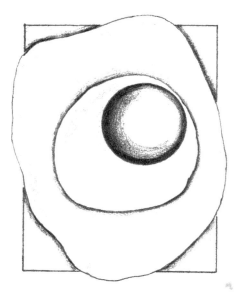

8. EGGS AND EGG PRODUCTS

Eggs provide an excellent source of protein, iron, vitamins A, D, E, and K and the B vitamin, riboflavin. However, the egg yolk contains a significant amount of cholesterol. The egg white contains no fat, is about 87% water, and contains more protein than the yolk.

USES OF EGGS

Eggs are used in many ways, alone as the main dish or in combination with other foods. Their unique characteristics give them a special place in food preparation.

Thickening Agent

Eggs are used to thicken a wide variety of food products, such as custards and sauces. Both the white and the yolk are capable of thickening when heated. This thickening process is called *coagulation*. Eggs coagulate over a wide range of temperatures, and personal judgment is necessary to recognize when the product has thickened to the desired end point.

When thickening egg products, the temperature must be watched very carefully. The mixture should be heated to *just below* the point where the eggs have reached maximum thickening, because they will continue to thicken for a few minutes after being removed from the heat. When this point is reached, the egg mixture should be served immediately or cooled. Overcooking may cause sauces to *curdle* (when the protein separates from the mixture), fried eggs to toughen, and custards to *weep* (when liquid drains from the custard upon standing). Plain eggs, such as fried or poached, coagulate much sooner than eggs that have been combined with liquids or have other ingredients added to them.

Binding Agent

The protein in eggs helps to bind ingredients and to keep the shape of products, such as meat loaf. Eggs are used to coat products in making crusts for breaded dishes by binding the crumbs to the products.

Emulsifying Agent

Eggs can also help hold food products together in other ways. As mentioned in Chapter 4, an *emulsion* is a dispersion of one liquid, into another. In mayonnaise, for example, the oil droplets are dispersed in liquid, and egg yolk is added to keep the oil from separating. The egg yolk attaches to the surfaces of the oil droplets and holds them in suspension so the oil will not separate from the liquid. Egg yolks work very well as emulsifiers because they are a combination of fat and water. They are a better emulsifying agent than whole eggs or egg whites. Cake batters depend on eggs to keep fat and liquid from separating.

Leavening Agent

Eggs are an important leavening agent when whipped to incorporate air. The egg white forms a *foam,* film around the air bubbles. When heated, the air bubbles expand and *leaven* the product (create a light product). As the heating continues the egg-white film coagulates and retains its expanded size, causing the product to remain light. Egg whites are used to leaven angel food cakes, sponge cakes, fluffy omelets and soufflés, and meringues.

Coloring and Flavoring Agent

When egg yolks are used in products, they add a richness in color as well as improve the flavor. Baked goods, noodles, cream fillings, and puddings gain their creamy color from egg yolks.

EGG QUALITY AND SIZE

Eggs are graded under the Federal-State Grading Program administered by the U.S. Department of Agriculture (USDA). *Grade* refers to the quality of the egg. Eggs are graded before storage and may not always reflect the egg quality when purchased because of improper handling and length of storage. Only grade A or AA should be used.

The color of the shell has no relation to quality of the egg. A high-quality egg will have a clean, uncracked shell, a yolk free from defects, and a firm white.

The weight of one dozen eggs provides an indication of general size of the eggs. The size of an individual egg within the dozen can vary somewhat because a larger egg can be offset by a smaller one. Standardized recipes used in quantity food production often are based on the use of medium eggs. *In preparation, eggs should be measured by weight rather than numbers of eggs because not all individual eggs are the same size.* Size and weight has no relationship to quality. A larger egg is not of higher quality than a smaller one.

Table 8.1 shows the weight of different sizes of unshelled eggs and number of eggs per quart. Six egg sizes have been established by the USDA, although only four are shown in the table and "Small" is not commonly available for purchase.

Table 8.1. Weight of different sizes of shell eggs and number per quart

Egg Size	Minimum Net Weight in Shell		Approximate Number per Quart (2 lb 2 oz)		
	1 doz (carton)	30 doz (case)	Whole eggs	Egg yolks	Egg whites
Extra large	27 oz (1 lb 11 oz)	50 lb 8 oz	17	49	26
Large	24 oz (1 lb 8 oz)	45 lb	19	55	29
Medium	21 oz (1 lb 5 oz)	39 lb 8 oz	22	63	33
Small	18 oz (1 lb 2 oz)	34 lb	25	74	39

Source: Adapted from USDA (1971).
Note: Size and grade are marked on the carton or case, but weight is not.

PURCHASING AND STORING EGGS

Purchasing

The following guidelines should be used when purchasing shell eggs (American Egg Board, 1991).

- Accept only clean, sound, odor-free eggs.
- Accept only eggs delivered under refrigeration or cooled to temperatures of 55 F.
- Accept only eggs packed in clean, snug-fitting fiberboard boxes.
- Check the grade of eggs delivered to ensure that they meet the specifications.

Storage

An egg can be stored for long periods of time, even up to several weeks if the conditions for storage are controlled carefully. The shell protects the egg, but the temperature and humidity must be controlled. Immediately after delivery, eggs should be stored under proper refrigeration of 45 F or below (do not freeze), with a relatively high humidity. Eggs lose quality fast when they are not chilled. The quality of a grade AA may drop to a grade B in a week if storage is at room temperature. Eggs also absorb odors, so they should be stored away from strong-smelling foods, such as onions, apples, and cabbage.

As eggs age or lose their quality, the thick portion of the egg white becomes thinner and the yolk moves away from the center of the egg and becomes flatter. A change in flavor also will occur.

Market Forms of Eggs

Eggs can be purchased as fresh shell eggs, dried whole-egg solids, egg-yolk or egg-white solids, or frozen whole-egg, egg-yolk, or egg-white solids. Processed eggs are convenient for quantity food production because they eliminate the time-consuming task of breaking eggs. With careful storage and preparation, whole dried or frozen eggs can be used for scrambled eggs and omelets. They often are used in mixes.

All frozen and dried egg products are inspected by the U.S. Department of Agriculture and pasteurized to destroy any bacteria that makes food poisonous. Frozen and dried eggs must be pasteurized and bear the USDA stamp. Pasteurized fluid eggs have a shelf life up to 14 days under refrigeration. Although pasteurization destroys salmonella, microorganisms that cause pasteurized eggs to spoil are not killed by pasteurization.

FROZEN EGG PRODUCTS. Frozen egg products provide consistent quality and avoid waste because whites and yolks can be bought separately when both are not needed (Table 8.2). Frozen eggs cannot be substituted for whole fresh eggs when fresh egg flavor is important to the product.

Table 8.2. Weight and volume measures for frozen eggs

Type of Eggs	Number of Eggs (large size)	Frozen Eggs Weight	Frozen Eggs Measure
Whole	9	1 lb	2 c less 2 TBSP
	10	1 lb 1¾ oz	2 c
	12	1 lb 5½ oz	2½ c
	25	2 lb 13 oz	1 qt 1¼ c
Yolks	10	6¼ oz	¾ c
	12	7½ oz	¾ c 2 TBSP
	26	1 lb	2 c less 2 TBSP
Whites	10	11½ oz	1¼ c 2 TBSP
	12	14 oz	1½ c 2 TBSP
	14	1 lb	2 c less 2 TBSP

Source: Adapted from USDA (1971).
Note: The same weight and volume measures may be used for shelled fresh eggs.

Frozen eggs must be stored at 0 F or lower and thawed in the cooler or refrigerator or under cold running water, never at room temperature. Only the amount needed should be thawed, and it should be stirred well before use. Because frozen eggs are perishable, they need to be used within 1–3 days of thawing.

Frozen egg whites can be stored in the freezer, thawed for use, and used in any way that fresh whites would be used. After freezing and thawing, the foaming ability and flavor are still excellent.

Plain thawed egg yolk tends to form a pasty, rubbery texture and does not mix well with other ingredients, such as sugar and milk. Frozen eggs are available with salt or sugar already added. This helps to make thawed egg yolks and whole eggs more acceptable in food preparation.

It is essential to use the right flavor for the product being prepared. For example, frozen whole eggs with sugar added will not be acceptable for scrambled eggs. (Frozen scrambled egg mixes are available for quantity foodservice use.) It also is important to know about any added of salt or sugar so adjustments can be made in recipes and foods can be selected correctly for modified diets.

REFRIGERATED LIQUID EGG PRODUCTS. Liquid egg products should be stored at 40 F or lower immediately upon delivery. The seal must be closed. Liquid egg products have a limited shelf life. They may be kept up to 6 days unopened, but once they are opened they should be used immediately.

DRIED EGG PRODUCTS. Dried eggs must be stored in the cooler or refrigerator at temperatures between 32 and 50 F and reconstituted right before use (Table 8.3). They must be tightly covered and refrigerated to prevent moisture and odor absorption and to prevent the formation of lumps. Objectionable changes in color, flavor and odor, and solubility occur when dried yolks or dried whole eggs are held at temperatures above 40 F. Emulsifying and foaming abilities of dried eggs may change, so the use of additives to improve those abilities may be required.

Table 8.3. Weight and volume measures for reconstituting dried eggs

Types of Eggs	Number of Eggs (large size)	Dried Eggs Weight	Dried Eggs Measure	Water
Whole	10	5 oz	1⅔ c	1⅔ c
	12	6 oz	2 c	2 c
	25	12½ oz	1 qt ¼ c	1 qt ¼ c
	32	1 lb	1 qt 1⅓ c	1 qt 1⅓ c
Yolks	10	3 oz	1¼ c	6⅔ TBSP
	12	3½ oz	1½ c	½ c
	54	1 lb	1 qt 2¾ c	2¼ c
Whites	10	1½ oz	6⅔ TBSP	1¼ c
	12	2 oz	½ c	1½ c
	100	1 lb	1 qt ¼ c	3 qt ½ c

Source: Adapted from USDA (1971).

Whole dried egg solids are very convenient to use. For breads, cookies, cakes, and similar recipes they are blended with the dry ingredients and the water required to reconstitute them is added to other liquids in the recipe. To combine dried eggs with dry ingredients in a recipe:

1. Weigh the dried eggs or measure them by placing them lightly in a measuring spoon or cup.
2. Blend them with the dry ingredients.
3. Calculate the water needed to reconstitute the dried eggs and add it to the liquid in the recipe.

For many types of recipes, the dried eggs are reconstituted before combining them with the rest of the ingredients. To reconstitute dried eggs with water, follow these steps:

1. Weigh the dried eggs or measure them by placing them lightly in a measuring spoon or cup, using exact weights or level measurements.
2. Sprinkle the dried eggs over the amount of water that is needed and blend by using a mixer, rotary beater, or wire whip.

EGG SUBSTITUTES. Several commercial egg substitutes are on the market in response to recommendations for cholesterol-free diets. These products usually do not contain yolk. Their color, flavor, and texture are generally found to be acceptable for scrambled eggs. These yolk-free products have more weeping than do whole-egg products.

Scratch cakes containing the egg substitutes will have lower volume and may not have as good flavor as those containing whole fresh eggs. On the other hand, the use of an egg substitute may result in a thicker stirred custard and a baked custard that has less sag.

FROZEN COOKED-EGG PRODUCTS. A frozen cooked-egg product is convenient and may or may not need heating after thawing. However, one problem with freezing cooked products is that the white becomes tough and rubbery. Weeping is a major problem for products that are cooked, frozen, thawed, and reheated.

COOKING EGGS

The principles of egg preparation are simple and unchanging. If they are followed, desirable products will result. Two are the most important:

- Avoid high temperatures
- Avoid long cooking times

The Food and Drug Administration (FDA) has developed the following precautions when handling eggs and egg-rich foods (American Egg Board, 1991):

- Eggs must be cooked thoroughly until both the yolk and white are firm and not runny, in order to kill any bacteria that may be present. Scrambled eggs need to be cooked until firm throughout.
- Adequate cooking times for different styles of eggs, which may have to be adjusted for egg size and initial temperature, include

  scrambled—1 minute at a cooking surface of 250 F
  poached—5 minutes in 2–4" boiling water
  fried—3 minutes at cooking surface of 250 F on one side and 2 minutes on the other
  soft-cooked—7 minutes in boiling water

- Scrambled eggs should be cooked in small batches (3 quarts or less) until there is no visible liquid egg, according to the rate of service.
- Serving lightly cooked foods containing shell eggs, such as stirred custard, meringues, or French toast, to people at high risk, such as the elderly or ill, may be unwise.
- Recipes and food-handling practices should use pasteurized egg products instead of shell eggs whenever possible.
- Raw or cooked eggs should not be kept out of the refrigerator for more than 1 hour, including the time for preparing and serving but not cooking.
- Cold egg dishes should be kept below 40 F and hot egg dishes above 140 F. Hot egg dishes should not be held on the serving line for more than 30 minutes.
- Eggs that have been held in a steam table pan should not be combined with a fresh batch of eggs; the fresh batch should instead be placed in a fresh pan.
- A raw egg mixture should not be added to a batch of cooked scrambled eggs being held on the steam table.

The temperature and length of heating affects the toughness of eggs. As the cooking temperature rises, the eggs toughen. The longer eggs are heated, the tougher they become. Therefore, eggs should be cooked at low to moderate temperatures for short periods of time. The new FDA guidelines require that the whole egg reach 140 F throughout for 3½ minutes.

Raw or unpasteurized eggs must be cooked completely to destroy the salmonella organisms. Raw eggs should not be used in egg-milk drinks, uncooked salad dressings, or when only partially cooked.

If eggs are broken into a separate bowl before using, the condition of the yolks and the presence of blood spots can be checked. In quantity food production, time and motion can be saved by breaking about 5 eggs into a separate bowl before putting them into the mixture. This procedure prevents an egg from spoiling more than 5 eggs at one time.

To prevent eggs from being cooked too quickly when they are being combined with hot liquid, a little hot liquid should be stirred into the beaten eggs to warm them. Then the eggs are stirred into the remaining hot liquid slowly while stirring continually.

Eggs should be brought to room temperature before using

- To prevent cracking if cooked in the shell
- To make the separation of white and yolk easier
- To hasten the beating process
- To increase the volume of beaten whole eggs, egg whites, or yolks

Methods of Cooking Eggs

COOKING UNSHELLED EGGS IN WATER. Eggs in the shell can be cooked in a pan on top of the range, in a wire basket in a steam-jacketed kettle, or in a compartment steamer. Before cooking, the eggs should be unrefrigerated only long enough to bring them to room temperature to prevent the shells from cracking when placed in warm water. Eggs with uncracked shells are cooked in boiling water. Hard-cooked eggs should be placed immediately in cold water to prevent formation of a green ring.

Here are cooking times based on room-temperature eggs:

Degree of Doneness	Time
Soft-cooked eggs	7 minutes
Hard-cooked eggs	16–20 minutes

A chilled egg will take a few minutes longer to cook than a room-temperature egg, and the smaller the egg, the less time is needed for cooking.

Hard-cooked eggs should have a tender, completely coagulated white with a mealy, firm-textured yolk with no discoloration. Soft-cooked eggs should have a tender, completely coagulated white with a yolk that is beginning to thicken. These characteristics are necessary to meet the necessary time and temperature requirements for safety as recommended by the FDA.

If eggs are to be diced or chopped in a quantity foodservice, cooking them out of the shell will save time and labor. Eggs are broken into a greased pan to a depth of 2 inches or less. They are cooked uncovered in a steamer for about 20 minutes. They also may be set over a pan of boiling water, covered, and baked in the oven for 40 minutes at 350 F. After the eggs are cooked, they should be cooled immediately and diced or chopped.

Hard-cooked, peeled eggs stored in a solution that keeps their quality are available for quantity foodservices. The appearance of these eggs may be less than perfect, but they are adequate for chopping up and adding to products such as salads.

SCRAMBLED EGGS. Before scrambled eggs are heated, the eggs are blended with a liquid, usually milk (1 tablespoon of milk or thin white sauce per egg or 1 cup per 16 eggs). Enough beating should be done to mix the eggs and liquid completely to an even yellow color. A foam should not result. Milk helps to make a more tender and moist product.

Eggs can be scrambled in a heated pan, a steam-jacketed kettle, a deep pan in the oven, or in a pan in a compartment steamer. Continuous stirring should occur during the entire cooking process, if possible. The stirring moves the cooked product from the bottom of the pan and allows the fluid egg to reach the heat. If cooked in the steamer, the scrambled eggs will remain in a solid mass that must be broken into pieces before serving.

A good scrambled egg has a shiny surface with medium-large, solid pieces and a uniform yellow color throughout. Scrambled eggs should be served immediately. If they are held too long they can turn an unappetizing green. Overcooking results in a firm, rubbery mass if little or no liquid is added. A curdlike mass may occur if too much liquid is added.

FRIED EGGS. A high-quality raw egg and controlled heating are necessary for a pleasing fried egg. Fried eggs should be prepared so the yolk is beginning to thicken (no longer runny, but not hard). The unbroken yolk should be covered by a film of coagulated white. The white should not have the crisp browning that results if the temperature is too high.

Eggs can be fried in a skillet on top of the range, in a tilting skillet, or in a pan and baked in the oven. Fried eggs are cooked in just enough fat to keep them from sticking. The white coating on the upper surface of the egg yolk should be coagulated. This can be done by basting the egg with hot fat from the pan, by adding a small amount of water and covering the pan, or by turning the egg over for a short period of time.

POACHED EGGS. Poaching is one method of egg preparation that helps decrease the amount of fat in the diet. When eggs are poached, the water should be heated to boiling before gently slipping the eggs into the pan. The action of boiling water, recommended by the American Egg Board, will not be as satisfactory as cooking in simmering water, but the eggs will be safer to eat. Only eggs of high quality will be satisfactory when cooked this way because they will stay together and not spread out as do lower-quality eggs when placed in water.

If the egg is broken into a separate bowl, its quality can be checked before it is placed in the water. For best results the egg should be slipped gently toward the side of the pan. A small amount of distilled vinegar or salt will help set whites and keep them from spreading.

The egg is removed with a slotted spoon as soon as the white has coagulated into a solid mass and the yolk is not runny, and then it is drained well. A good poached egg should have a firm, tender white piling around a slightly thickened unbroken yolk.

Eggs can be poached in a shallow pan on top of the range, in the oven, in a poaching pan, or in a steamer. In quantity food production, a large shallow pan is used with a continuous heat source to distribute heat evenly to all parts of the pan. To poach eggs in advance, the eggs can be removed when barely done and held in ice water. At serving time the eggs are reheated for 30–60 seconds in simmering liquid.

BAKED EGGS. Baking eggs requires little or no fat and is often used as a substitute for poaching eggs in quantity foodservice. Eggs are broken into individual pans or a large, shallow, greased pan and baked at 325 F until the desired doneness. Butter or margarine, milk, salt, and pepper may be added to the top before baking, but it is not essential.

Baked eggs should have a tender white that is completely coagulated, with the yolk thickened but not set. There should be no signs of browning around the edges.

SHIRRED EGGS. *Shirred eggs* are baked or broiled eggs that are fried on top of the range until the white begins to firm and the edges turn white (about 1 minute). Then they are placed under the broiler for 2–4 minutes or baked in a moderate oven (350 F) to cook the top.

OMELETS. Omelets are made with a mixture similar to scrambled eggs, but they are left in a single mass rather than in large pieces. In preparing an omelet, the mixture of eggs, milk, and seasonings are blended well and cooked at a moderate temperature. Adding 2 or 3 drops of water before beating makes a lighter, fluffier omelet. As the bottom of the omelet coagulates, a small portion of the egg mixture is lifted slightly to allow liquid egg to reach the heat. Omelets can also be cooked on sheet pans and cut into portions.

Ingredients, such as cheese, can be added and cooked as a part of the omelet. Any precooked mixture of ingredients should be added before folding the omelet. Sauces can be spooned over the omelet. A perfect omelet should be fluffy, moist, and tender; firm throughout; golden brown in color with no dark brown spots; oval in shape; and in one continuous piece.

SOUFFLÉS. A *soufflé*, which means "puffed," is similar to an omelet. However, a soufflé has a yolk mixture that is actually a thick white sauce in which egg whites are incorporated. The soufflé is baked in an oven in water to slow down the transfer of heat reaching the soufflé. When the soufflé is done, less shrinkage will occur if it is left in an open oven for a few minutes.

Some soufflés contain strained vegetables and/or ground meats and are served as a main course for lunch or a side dish for dinner. Soufflés also may contain grated cheese, fruit, or chocolate.

A soufflé should be high and light, pleasingly browned, and well blended with no suggestions of layers. It should be tender and flavorful.

CUSTARDS. *Custards* are sweetened milk mixtures thickened with egg that usually have salt and vanilla added. Custards can be either soft or baked. A soft custard, which results from continuous stirring of the mixture, is done when the mixture coats the spoon, a test that involves personal judgment. It is important to remove the custard immediately from the heat and cool it rapidly by pouring it into a shallow pan with ice around it to stop any further cooking. Soft custards are less likely to curdle if they are cooked above water that is just below the boiling point.

Baked custards are cooked in the oven without any stirring. They cannot be cooled as quickly as soft custards because baked custards are gelled. Because the temperature in the center of the custard will rise even after removing it from the oven, it should be removed before the center has coagulated completely. To test for doneness, a clean, stainless steel or silver knife is inserted halfway between the edge and center. If the custard does not cling to the knife when it is withdrawn, the custard is ready to remove from the oven so it will not become overcoagulated and weep.

In both soft and baked custards, eggs provide most of the thickening. The salt in the milk or as a separate ingredient provides the mineral needed for the eggs to coagulate.

Foams

Foams are dispersions in which bubbles of air are surrounded by thin layers of liquid. Egg white can increase its surface area around the gas bubble without squeezing out the air too quickly. Eggs at room temperature can be beaten more easily into a foam than can colder ones, and the stability and volume of the foam will also increase. A longer time is required for foam formation if ingredients such as water, fat, salt, sugar, and egg yolk are present.

EGG-YOLK FOAMS. Yolk foams are used in certain baked products, such as sponge cake and puffy omelets. Egg yolks are beaten into heavy foams, which increases in the volume and lightens the color. The stability of the yolk foam is just enough to allow it to be folded with other ingredients and baked. If salt is omitted from a recipe that uses yolk foam, the product will have a lower volume and be less tender.

Yolk foams pile when they are beaten for a long period of time, but they never become stiff enough to peak. In quantity recipes this time is about 10–15 minutes. Whole eggs can also be beaten to a soft foam.

EGG-WHITE FOAMS. Egg whites are very useful when a foam is needed. Egg white foams are more stable and elastic than yolk or whole-egg foams. With mixing, air can be incorporated into the bubbles created by beating the proteins.

STABILITY AND VOLUME. The two important factors in egg-white foams are stability and volume. As beating time increases, the stability and volume of the egg-white foam increases and the surface looks dry. Maximum stability is reached before maximum volume. As maximum volume occurs, the bubbles become very small and less stable. This stage is called the dry-peak stage. The egg-white foam has maximum stability at the stiff-peak stage.

Ingredients such as sugar, cream of tartar, water, salt, or fat affect the egg-white foam. The addition of cream of tartar makes the foam more acidic, increasing the stability, which results in less shrinkage during cooking. For every 2 pounds of egg white, 1 teaspoon of cream of tartar should be added. Water will greatly increase the volume of egg-white foam, while decreasing stability, which results in a very poor-quality product.

Any added salt reduces the stability of the foam. This problem might be overcome by greatly increasing the beating time. The addition of sugar to the egg whites makes a foam that is less stiff, more plastic, and more stable than foams containing no sugar. However, adding sugar greatly increases the beating time. The highest-quality sugar-containing foam is created by beating the whites until they hold soft peaks and then adding sugar slowly while continuing to beat.

If egg yolk or any fat gets into egg whites, both stability and volume are decreased. The procedure of separating 5 eggs at a time into a separate bowl will prevent wasting many eggs if any yolk does get into the whites.

The problem of yolk contamination is impossible to avoid in frozen or dried egg whites. This problem can be solved by adding additional freeze-dried egg white to liquid whites or increasing the total quantity of egg whites used.

MERINGUES. *Meringues* are egg-white foams with sugar added. Soft meringues, the type used on cream pies, usually have 2 tablespoons of sugar per egg white. A well-prepared meringue will have a good volume, fine texture, light-colored to golden brown surface without any moisture, and no weeping.

Weeping can be prevented by putting the meringue on the filling while the filling is still very hot and baking immediately. The heat from the pie will help bring the temperature of the meringue near the surface of the pie to a temperature high enough to coagulate the egg white. When the meringue is placed on a hot filling (140–170 F), however, the baking time must be shortened to keep the meringue from being overcooked. Overcoagulation of the egg white on the surface of the meringue will result in *beading,* the collection of small, amberlike drops on the surface of the baked meringue. Overcooking is also likely to result in a meringue that is sticky and difficult to cut.

When a meringue is placed on a filling that is too cool (below 140 F) or too hot (above 170 F), the exterior of the meringue will overcook while the interior will be undercooked and therefore unsafe to serve. If the filling is above 170 F, the baking time may need to be reduced. It is important that meringues are cooked above 140 F throughout.

SUMMARY

Eggs provide an excellent source of protein, iron, and some vitamins. They are used in food to thicken, emulsify, leaven, bind, coat, and add color and flavor. Market forms of eggs include fresh, dried, and frozen. Frozen cooked-egg products and egg substitutes also are available. The planned use of eggs in the menu needs to be considered when selecting the form to purchase. Proper storage of eggs is important because the quality of the egg deteriorates quickly.

The main principles of egg cooking are to avoid high temperatures and long cooking times. The safety of eggs in the product must always be considered. Various methods of cooking can be used to add variety to the cycle menus. Each method has additional preparation principles.

Eggs are also a major ingredient in foams, which are used for baked products, desserts, and omelets. Methods that increase stability and get the right volume must be used.

LEARNING ACTIVITIES

Activity 1: Comparing Two Methods of Cooling Hard-cooked Eggs

1. Prepare 2 hard-cooked eggs. Remove them from the hot water.

2. Place 1 egg immediately in cold water to cool. Let the other one cool at room temperature.

3. Peel both eggs and compare them.

    A. Did the eggs differ in appearance?

    _____

    B. Was there any difference in flavor? If so, what?

    _____

    C. Which did you prefer? Why?

    _____

Activity 2: Comparing Two Methods of Storing Eggs before Cooking

1. Store 2 eggs for 1–2 weeks: one at room temperature and one in the refrigerator.

2. Observe the raw eggs.

    A. Carefully crack each egg into a flat dish.

    B. Observe the appearance of the yolks and whites.

    C. Compare the sizes of the egg whites at the widest point.

3. Fry both eggs.

    A. Melt about ½ teaspoon of butter or margarine in a skillet over low heat until it sizzles. Be sure to coat the entire bottom.

    B. Carefully slip 1 egg from its dish into the skillet.

    C. Cook until the egg white is firm around the edge, and then add 1–2 tablespoons of water to the skillet, around the egg.

    D. Cover the skillet tightly and cook 3 minutes, until the egg white is completely coagulated.

    E. Baste with water several times.

    F. Observe the appearance of the egg and the size of the egg white.

    G. Fry the second egg, using the same procedure.

4. Compare the cooked eggs.

   A. Was either egg yolk not centered? Why not?

   _____

   _____

   B. What is the relationship between the appearance of the egg whites and the amount of spread?

   _____

   _____

Activity 3: Comparing Two Methods of Cooking Scrambled Eggs

1. Prepare two servings of scrambled eggs using the same recipe for each, one in a frying pan and the other in the top of a double boiler.

2. Observe the cooked eggs after the times shown in the chart and record the appearance each time.

Activity 3, Step 2

	Scrambled Eggs after Standing		
	10 min	30 min	40 min
In frying pan			
In double boiler			

Activity 4: Comparing Methods of Preparing Egg Foams

1. Blend 12 tablespoons of egg whites.

2. Place 2 tablespoons of the blended egg whites into each of six 2-cup measures. Beat each cup of egg whites until foamy.

3. Add one of the following ingredients to each 2 tablespoons of egg whites.

   A. None

   B. 1 tablespoon of tap water

   C. 2 tablespoons of sugar

   D. ½ teaspoon of egg yolk

E. ⅛ teaspoon of cream of tartar

F. ¼ teaspoon of salt

4. Beat each sample of egg whites until the peaks are stiff but not dry and hold their shapes, but do not overbeat. Record the beating time of each.

5. Observe the differences in appearance of each sample.

6. Level the beaten egg whites with a spatula and record the volume of each.

7. Let the samples stand for 15 minutes.

8. Repeat step 5.

9. Which treatment produced the largest volume? The least volume? The most stable product?

_____

What are the reasons for these differences?

_____

_____

_____

Activity 5: Comparing the Quality of Custards with Different Cooking Times

1. Prepare custard following these steps:

    A. Beat 1 egg just enough to blend the yolk and white. Do not allow a foam to form.

    B. Add 2 tablespoons of sugar and ½ teaspoon of salt. Blend.

    C. Add 1 cup of milk and ⅛ teaspoon of vanilla and strain.

2. Cook over simmering water (185–195 F) until the mixture coats a spoon.

3. Remove from the heat and record the time it cooked.

4. Pour out ⅓ of the mixture.

5. Continue cooking until mixture is thick.

6. Remove from the heat and record the time the remaining ⅔ cooked.

7. Pour out ⅓ of the mixture.

8. Continue cooking until the mixture is curdled.

9. Compare the three samples and fill in the chart.

Activity 5, Step 9

Sample No.	Time Cooked	Thickness (slightly thickened, thickened when cooled, watery, runny thick)	Texture (smooth, curdled, lumpy)	Flavor (good, flat, tasteless)
1				
2				
3				

Activity 6: The Functions of Eggs in Cooking

1. Complete the chart by giving examples for use in food preparation for each of the functions listed.

Activity 6, Step 1

Function	Example
Emulsify	
Leaven	
Thicken	
Flavor/color	

Activity 7: Comparing Eggs Hard-cooked by Two Methods

1. Hard-cook 12 eggs out of the shell in a pan in the steamer. Hard-cook 12 eggs in boiling water in a steam-jacketed kettle or in a pan on top of the range. Follow the recipe for both methods.

2. Peel and chop the eggs.

    A. Compare the quality of the eggs for flavor, texture, and color.

    B. Compare the preparation times for the two methods.

Hard-cooked Eggs		
Ingredients	**Amount**	**Procedure**

Steam-jacketed kettle (in shell)
Eggs, shell ....
Water, boiling ....
	As needed
	To cover
		1. Place eggs in steam-jacketed kettle wire basket.
		2. Fill kettle with enough water to cover basket of eggs; heat to boiling.
		3. Lower basket of eggs into boiling water. Adjust pressure so water simmers.
		   Note: DO NOT BOIL.
		4. Cook 12–15 min.
		   Note: Test an egg for doneness before removing basket.
		5. Remove basket of eggs and chill immediately in cold running water.
		6. Remove shells. Crack shells by tapping eggs against a hard surface, then loosen by rolling on hard surface; start peeling at large end. (If membrane adheres to white and egg is difficult to peel, the egg was too cold or too fresh.)

Steamer (out of shell) Eggs, shell	As needed	1. Crack eggs into 12×20×2½" solid, greased counter pan. (3–3½ doz per 12×20×2½" pan maximum. DO NOT USE DEEP PANS.) 2. Place pans of eggs in preheated steamer and cook. *Time at 5-lb pressure:* 6–8 min 3. Check for doneness by piercing yolk in center of pan. 4. Remove from steamer and drain off any accumulated condensation. Chill immediately in pan of ice or chop immediately to prevent continued cooking.

Source: Adapted from *Standardized Quantity Recipe File* (1971).

Activity 8: Comparing Eggs Scrambled with Various Liquids

1. Scramble four batches of 12 eggs as follows:

    A. Scramble one batch without liquid.

    B. Scramble one batch with ¾ cup of liquid (1 tablespoon per egg).

    C. Scramble one batch with a thin white sauce as the liquid.

2. What differences in quality of the eggs did you notice for flavor, texture, consistency, and color?

___

___

Activity 9: Comparing Various Custards

1. Divide into smaller groups. Each group cooks one of the following custards using "home-quantity" recipes.

    A. Baked custard using frozen eggs and dried milk

    B. Baked custard using fresh eggs and milk

    C. Baked custard without salt

    D. Stirred custard with frozen eggs and dried milk

    E. Stirred custard with fresh eggs and milk

    F. Stirred custard without salt

2. What differences did you note in the quality of the custards for flavor, texture, color, and consistency? Fill in the chart.

Activity 9, Step 2

	Flavor	Texture	Color	Consistency
A				
B				
C				
D				
E				
F				

3. What effect was there when salt was left out of the stirred custard? The baked custard?

___

___

REVIEW QUESTIONS

True or False

1. An egg will drop to a lower grade if storage conditions are not good.

    A. True
    B. False

2. The purchaser can be assured that the quality grade on the egg carton reflects the grade when purchased.

    A. True
    B. False

Multiple Choice

3. The recommended temperature and time to use in cooking eggs is

    A. High temperature for a short time
    B. Low temperature for a short time
    C. High temperature for a long time
    D. Low temperature for a long time

4. Hard-cooked eggs in the shell should be placed immediately into cold water to

    A. Keep the whites from toughening before the heat reaches the yolk
    B. Reduce the degree of temperature change between refrigeration and hot water
    C. Prevent a green ring from forming around the yolk
    D. All of the above

5. To increase the stability of egg-white foams

    A. Make sure the whites are as thin as possible
    B. Add sugar, cream of tartar, or lemon juice
    C. Have the eggs at room temperature
    D. Beat the eggs only until they are double in volume

6. The quality of eggs is determined by

    A. The condition of the yolk
    B. The weight of the egg
    C. The firmness of the white
    D. The color of the shell
    E. Both A and C
    F. All of the above

7. When egg yolk is used to keep one liquid dispersed into another, it is called

   A. Leavening
   B. Emulsifying
   C. Coagulating
   D. Binding

8. A dozen eggs labeled "large" means what?

   A. All the eggs are at least size large.
   B. All eggs will be higher quality than medium.
   C. The dozen eggs have that general egg size.
   D. There cannot be any small eggs in the dozen.

9. As eggs lose their quality, what happens?

   A. The egg white gets thicker.
   B. The shell will thicken.
   C. The yolk will develop blood spots.
   D. The yolk moves away from the center of the egg.

9. DOUGHS, BATTERS, AND PASTRIES

High-quality bakery products are a popular part of meals. These products provide important nutrients and calories, as well as some of the fiber necessary for a well-balanced diet.

Yeast breads, quick breads, batters, pastries, cakes, and cookies have similar ingredients. However, the preparation methods are different for each.

BREAD DOUGHS AND BATTERS

Yeast Breads

Yeast breads are baked dough that is usually made from flour, water, salt, sugar, and yeast. An understanding of yeast fermentation and gluten development are necessary before beginning to make these products in quantity.

To produce a high-quality yeast dough, the amount, type, and quality of ingredients must be carefully controlled. For example, the baker should be aware that the gluten content of hotel and restaurant flour may vary from one brand of flour to another. Even the same brand may vary from one purchase to another. Whenever possible, flours of similar gluten content should be used. Consistent quality can be assured by specifying the percentage of protein in the flour on purchase specifications.

Bread flour is made from varieties of wheat containing proteins that produce very strong gluten. All-purpose flour and cake flours are made from varieties with less gluten potential. The milling process also influences the amount of protein in the final product, so different flours can be made from the same variety of wheat. The milling process separates the *wheat germ* (the interior fat-containing kernel), the starchy layer, and the external bran. The *bran* (the husk covering of the wheat) is the primary source of fiber in all types of flour.

Gluten is responsible for most of the structure in bakery products. It is traps leavening gases and consequently affects the volume of the final product. Too much gluten makes a tough final product and too little gluten results in a weak, fallen product.

TYPES OF FLOUR. Using the best flour for the product results in the best quality with controlled costs. Here is a list of some of the types of flour that can be purchased and their specific uses:

Types of Flour	Uses
Straight flour	White breads, whole wheat–blend breads
All-purpose flour	Cakes, cookies, quick breads
Bread and patent flour	White breads, whole wheat–blend breads
Clear flour	Rye breads
Cake flour	Cakes and delicate baked goods
Pastry flour	Pie doughs, cookies, biscuits, muffins
Whole wheat flour	Whole wheat breads
Bran flour	High-fiber breads
Self-rising flour	Breads, some pastries

Whole Wheat Flour. Whole wheat flour produces breads that are high in fiber content. Whole wheat flour, also called graham flour, is made by grinding the entire wheat grain, including the bran and germ. Because the wheat germ is included, this flour becomes rancid very easily and has a shorter shelf life than does bread or all-purpose flour. However, refrigeration will extend the life of the flour.

Whole wheat flour is suitable for yeast breads because it contains gluten proteins. Bread made solely from 100% whole wheat flour will be much heavier than white bread. This heavy, dense structure forms because the sharp edges of the bran cut the gluten strands. In addition, as the bran fiber increases, proteins decrease, because bran does not contain gluten proteins, and this results in a lower volume.

The fat from the wheat germ contributes to the shortening action of whole wheat flour because the oil in wheat germ also acts as a tenderizing agent. Other fats, oils, and sugars also help tenderize the product.

Whole Wheat Blends. To assure a larger loaf, many commercial whole wheat breads also contain white bread flour. If whole wheat flour is not available, miller's bran can be added to white flour to produce whole grain bread. With this substitution, adjustments in the quantity of water used may be necessary.

Rye Flour. Although it does not contain all the proteins required for gluten, rye is also a popular flour used in yeast breads. Breads made from 100% rye flour will be very heavy and dense. Bread flour or all-purpose flour must be added to the dough if a lighter loaf is wanted. Recommended ratios are 25–40% hard wheat flour to 60–75% rye flour. Here are two examples of these proportions for a 5-pound yeast bread:

	Rye Flour	Hard Wheat Flour
Recipe A	3 lb	1 lb
Recipe B	2.4 lb	1.6 lb

Rye flour is used to make rye and pumpernickel breads and similar specialty products.

DOUGHS, BATTERS, AND PASTRIES

MIXING YEAST BREAD DOUGH. To make a high-quality bread, it is important to use the correct mixing technique. There are three important mixing procedures for all methods of preparing yeast dough.

- Blend all ingredients evenly.
- Develop the gluten for structure and smooth appearance.
- Distribute the yeast evenly throughout the dough.

If gluten is overdeveloped as the result of improper handling and/or overmixing, the final product will be tough. Foods like muffins will have large, oval-shaped holes called tunnels. Yeast doughs are prepared using one of three methods: straight-dough, modified-straight-dough, and sponge.

Straight-Dough Method. The straight-dough method is the simplest. The yeast is softened in a little water before it is added to the other ingredients. Then the ingredients are placed in a mixing bowl and mixed in one step, using a dough hook attachment. This method produces a high-quality plain white bread.

Modified-Straight-Dough Method. The modified-straight-dough method is used primarily for rich, sweet doughs. Like the straight-dough method, the yeast is presoftened. The fat, dry milk solids, salt, and sugar are mixed together with a whip until they are evenly combined. The eggs are added slowly to the mixture to ensure even distribution. The liquid is added all at once and then the flour and yeast. The dough is mixed with the dough hook until it is smooth and cleans the sides of the bowl.

Sponge Method. The sponge method mixes the dough in two stages to give the yeast extra time to leaven the bread. This method is best for breads with lower gluten, such as rye, whole grain, and other breads containing the whole-grain kernel. First, the liquid, yeast, and part of the flour are mixed to form a soft dough. This dough is allowed to expand to double in bulk before being punched down. The remaining flour and ingredients are then added and the dough is kneaded using a dough hook until it has a smooth texture.

Milk and Salt in Yeast Doughs. To increase the yeast dough volume in all three methods, milk should be *scalded* (heated to 187 F). Scalding the milk kills any microorganisms that might prevent the yeast from growing. The milk proteins begin to coagulate during scalding, allowing them to help build structure in the bread. Even when dry milk is reconstituted in the modified-straight-dough method, the milk should be scalded. Scalded fresh milk can be substituted for the scalded reconstituted dry milk and added to the dough with the other liquids. In quantity food production, the advantages of the extra volume possible when milk is scalded should be compared with the efficiency by eliminating the scalding process.

Salt also must be carefully measured for all yeast bread recipes. Salt not only adds flavor, but is also a major control of the yeast growth. It cannot be reduced in an already established recipe without causing excessive yeast growth.

PROOFING YEAST BREADS. The *proofing* (rising) that occurs in yeast doughs is the result of *yeast*, the leavening agent. Yeast is available as a cake or as active dry granules. Yeast ferments the sugar in the recipe to form carbon dioxide, alcohol, and water. The sucrase enzyme produced by the yeast breaks down any *sucrose* (table sugar) in the formulation into *fructose* and *glucose*, which are simple sugars that the yeast can ferment. When the recipe does not have added sugar, the yeast will use *glucose* (sugar that is released from the starch in the flour). Because the glucose must be released by an enzyme present in the flour, doughs without added sugar take longer to proof.

The carbon dioxide remains in the product until it is baked, forming the necessary pockets of air that leaven and proof the bread. This process is called *proofing* in yeast breads and *leavening* in quick breads. As more and more carbon dioxide is formed, the pockets grow larger. As the carbon dioxide heats in the oven, it expands and exerts pressure that causes the dough to expand. The alcohol produced by fermentation evaporates during baking.

Most quantity-foodservice operations use a proofing cabinet to produce the heat and moisture needed to develop the yeast. Without the proper conditions and time for proofing, it is difficult to produce a flavorful bread of good texture and volume. The desired temperature for proofing is 85–95 F. Higher temperatures will shorten proofing time but will diminish the quality of the flavor and texture. Caution should be taken to ensure that the proofing temperature never exceeds 110 F. Otherwise, the yeast will be killed and the bread will not rise to full volume.

BAKING YEAST BREADS. Baking equipment and utensils should be selected for their intended use.

The recipe directions include the correct pan to use. During baking, pans should not be overlapped nor should they be placed directly over each other. This allows even distribution of the heat from one oven rack to another. Although oven temperatures eventually kill yeast, yeast produces additional carbon dioxide until the internal temperature reaches 110 F. The increase in volume that occurs in the oven is called *oven spring*. Preheating controls the amount of oven spring and ensures the yeast is killed before the outer crust is baked. Where this does not happen, the bread continues to proof and the crust splits.

When the oven is loaded, the door should be open the minimum time possible because the oven temperature cools every time pans are added. In addition, if too many pans are put in at one time, the temperature may cool off too much. Any such cooling causes the bread to overproof, be coarse in texture, and split.

Due to the delicate structure of baked products, the oven should not be opened to check for doneness until close to the end of the expected baking time. Before this, the tender and weak state of the bread could cause the product to fall. Yeast breads are usually done when they are golden brown, produce a hollow sound when tapped, or spring back into place when the product is pushed with a finger.

PROBLEMS WITH YEAST BREADS. Here are the causes for unsatisfactory bread products:

Problem	Causes
Volume	
Reduced volume	Too much salt
	Too little yeast
	Not enough liquid
	Low-gluten flour
	Under- or overmixing
	Too hot oven
	Poor-quality yeast
	Too little proofing
	Too much sugar
Too much volume	Too little salt
	Too much yeast
	Too much time elapsed before dough is panned and baked
	Overproofing

Problem	Causes
Poor shape	Too much liquid
	Flour with little gluten-producing protein
	Improper molding
	Improper fermentation or proofing
	Too much steam in oven
Split or burst crust	Overmixing
	Underfermentation
	Improper shaping of loaf; seam not on bottom
	Uneven heat in oven
	Too hot oven
	Insufficient steam
Texture and crumb	
Too dense or close-grained	Too much or too little salt
	Too little liquid
	Too little yeast
	Underfermentation
	Underproofing
Too coarse or open-grained	Too much yeast (open-grained)
	Too much liquid (coarse)
	Overmixing
	Overfermentation
	Overproofing
	Too large pan
	Undermixing
Streaked crumb	Improper mixing
	Poor shaping
	Too much flour used for dusting
Poor texture or crumbly	Too weak flour
	Too little salt
	Too long or too short fermentation time
	Overproofing
	Too low baking temperature
Gray crumb	Too high fermentation time or temperature
	Poor flour quality
Crust	
Too dark	Too much sugar or milk
	Underfermentation
	Overproofing
	Too high oven temperature
	Too long baking time
	Too much steam in oven

Problem	Causes
Too thick	Too little sugar or fat
	Oven not preheated
	Improper fermentation
	Baking too long and/or at wrong temperature
	Too little steam
Blisters on crust	Too much liquid
	Improper fermentation
	Improper shaping of loaf
Flavor	
Flat taste	Too little salt
Poor flavor	Inferior, spoiled, or rancid ingredients
	Poor bakeshop sanitation
	Under- or overfermentation
	Unclean baking pans

Quick Breads

Quick breads are much easier and less labor-intensive to produce than yeast breads. There are many versions, depending on the type of flour and other ingredients. Nuts, bran, spices, and fruits can be added to a basic recipe for variety and to increase fiber in the menu. Many quantity foodservices use quick bread mixes to reduce production time, save on expensive labor, and ensure standardized quality in the final product.

The primary agents that cause increase in volume (leavening) in these quick breads are baking soda and baking powder. Air and steam produced by the liquid in the recipe also contribute to the product volume. Baking soda (sodium bicarbonate) and acid from other ingredients react to form carbon dioxide. Common acid ingredients are sour milk or cream, chocolate, fruit or fruit juice, and molasses. Baking powder contains baking soda, a dried or crystalline acid ingredient, and starch. Baking powders are classified as single- or double-acting. Single-acting baking powder reacts at room temperature, so it may lose too much carbon dioxide as the batter is mixed. With double-acting baking powder, part of the reaction requires heat, so not as much carbon dioxide is lost during mixing. Double-acting baking powder is preferred for quantity food production to ensure proper leavening. Single-acting baking powder is rare and sometimes difficult to find, but it may be specified in older or ethnic recipes.

The mixtures used to make quick breads are usually soft doughs or batters. The soft doughs are used for rolled and drop biscuits, and batters are used for quick breads such as muffins and pancakes.

As in yeast dough, the main structural ingredient in quick bread is the flour. The starch and protein add structure to the breads. Because only slight development of gluten is desired in quick-bread products, the dough or batter is handled less during mixing. The amount of fat and sugar in the recipe also help to prevent overdevelopment of the gluten and contribute to tenderness and flavor. A recipe with a large amount of sugar and/or fat requires more mixing than one with very little.

Eggs are another important ingredient in quick breads. They contribute to structure, and beating them also helps incorporate air in the mixture. By the stretching of egg-white cells to trap more air, which turns to steam when heated, they help the product to retain the steam produced by baking.

QUICK-BREAD MIXES. Many quantity foodservices choose to use a variety of quick-bread and muffin institutional mixes. Because many of these mixes have dried egg and fat solids, only liquid is added during mixing. Most mixes have higher levels of sugar and fat than do products made from scratch. These additions add tenderness and longer shelf life to the final product and prevent gluten overdevelopment, but they usually have a finer cakelike texture than do recipes made from scratch.

QUICK-BREAD MIXING METHODS. There are three basic methods for mixing quick breads and muffins: biscuit, muffin, and creaming or cake. The mixing method is important in determining the expected quality, tenderness, volume, and texture of the final product. Each method contributes different characteristics to quick breads.

Biscuit Method. In the biscuit method, also known as the pastry method, the dry ingredients are thoroughly combined first. Next, the fat is cut into the dry ingredients with a flat beater attachment until the fat is broken down into pea-size clumps evenly distributed throughout the product. These clumps of fat allow for pockets to be formed that catch air and carbon dioxide released from the baking powder. Finally, the liquid is added.

Caution should be taken when mixing biscuit ingredients. Overmixing gives a fine texture, much like cake, and undermixing results in a product with a coarse texture.

After mixing, some kneading of biscuit products on a lightly floured surface will develop layers of fat in the dough and some gluten. Because gluten is easily developed, it is best to knead softly with the fingertips to work flour on the table into the dough. This light kneading produces a light, flaky product.

Muffin Method. In the muffin method, the dry ingredients are mixed together first to ensure an even distribution. The beaten eggs, milk, and oil are mixed together and then added all at once. This method is often used in recipes with minimal tenderizing ingredients (fat and sugar) and so as little mixing as possible is done, to prevent overdevelopment of the gluten that would result in the loss of volume and the formation of tunnels in the final product. However, some recipes with *high* proportions of sugar or fat call for this method and may require some stirring, for example, carrot cakes.

With the muffin method, overmixing can also occur in the dipping or portioning of the batter. A scoop that holds exactly one portion should be used in order to prevent the need for extra dipping. It is best to make several small batches rather than one large batch. Quantity foodservices should not prepare batches larger than 100 portions at a time to prevent the portioning itself from overmixing the batter. With small batches, the dry and liquid ingredients can be premeasured and then combined before portioning into the baking pan.

Cake and Creaming Method. The cake method, also called the creaming method, combines and *creams* (beats to a creamy texture) the sugar and fat to incorporate air into the product. Beaten eggs are added next, which incorporates more air. Finally, the liquid and the mixed sifted ingredients are added alternately. An alternative to this step is the *quick-mix method* when the dry and liquid ingredients are quickly added together and mixed, excluding a step for creaming. It is used in many quantity foodservice operations to save production time.

PANNING AND BAKING. Muffin tins and baking pans should be greased with a food-release product and sometimes recipes require a light dusting with flour as well. Paper muffin liners also can be used. They make the pans easier to wash, but they add to the cost and often are not well accepted by clients. Unwrapping food is difficult for some clients and unappetizing to others. Sheet pans used for products like corn bread can be greased lightly or lined with bun pan paper liners.

Baking quick breads follows many of the same rules as for yeast breads. A preheated convection or conventional oven is essential. A convection oven takes much less time. A conventional oven should not be overloaded, and one pan should not be placed directly above another, to allow for even distribution of the heat.

Because of the delicate structure of quick breads, the oven door shouldn't be opened to check for doneness until close to the end of the expected baking time. Opening it too soon could result in the product falling. This reduction of batter height and volume is caused by vibration, or movement, of the soft, partially cooked batter. As the batter moves, the pockets of leavening gases break open like burst balloons, the gas escapes, and the batter deflates.

Quick breads are tested by piercing them with a plain, uncolored toothpick or a metal cake tester. If no batter clings to the tester, the product is done.

PANCAKE AND WAFFLE BATTERS

Baking powder is used as the leavening agent in pancake and waffle batters. Most mixes require only the addition of water and sometimes buttermilk.

The use of an institutional pancake dispenser or proper-size dipper for pancake and waffle batter will help standardize portions in quantity foodservice operations. The grill should be heated to 350 F. If the temperature is too hot, the final product may be dark on the outside but still not done inside. A cold grill will cause the product to be pale and have a poor, dense structure. The pancake is ready to turn when bubbles form on the surface. The best products result when turned only once.

CAKES AND COOKIES

One of the most popular items on any menu is the dessert. With the sweetness and richness of desserts, however, also come the calories associated with fat and sugar. Some modifications in traditional cakes and cookies can reduce the fat and sugar, and thus the calories, while still allowing variety and flavor.

Cakes

Cakes are usually classified into two categories: foam cakes and shortened cakes. Foam cakes include angel food cake, sponge cake, and chiffon cake. Layer cake and pound cake are types of shortened cakes.

CAKE INGREDIENTS. The basic ingredients for cakes include flour, sugar, eggs, fat, water, and a leavening agent, such as baking powder or baking soda. Because water and fat do not mix naturally, an emulsion needs to be formed during mixing.

Beaten eggs and cake flour are the major structure-forming ingredients in cake. Too much sugar will result in a small or collapsed cake. Sugar is a tenderizer and may cause the cake to fall. In addition, extra sugar uses up the water needed to form the structure that holds up the cake.

Fats and oils tenderize bakery products by preventing the development of the gluten, and they may also add flavor. Emulsifiers in some shortenings permit cakes to be made with higher levels of water and sugar. These fats, called *high-ratio shortenings,* allow the cake to be moister and more tender than would be possible with shortenings that are not emulsified. This term *high-ratio* refers to the large amount of sugar and liquid in relation to the amount of fat. Regular shortening, butter, and margarine are not equal substitutes for a high-ratio, emulsified shortening.

Eggs have many functions in bakery products. Beaten eggs incorporate air and volume into a mixture. They also help products retain the steam released by heat. Eggs add structure, flavor, and color to the product. A large amount of eggs in a cake recipe, however, can produce toughness.

In egg-white-foam cake recipes, cream of tartar usually is used to stabilize the beaten egg whites and keep them from breaking or losing their volume after whipping. Cream of tartar functions by increasing the acidity of the egg-white mixture. It also helps to create smaller air cells in the mixture, producing a finer texture in the cake. When egg yolks are included in a shortened cake, they act as emulsifiers. Because the egg yolk is a combination of fat and water, it easily blends with both in a cake batter. The yolk provides a way for the usually unblendable oil and water to mix together in a smooth batter.

Cake flour will produce a cake with a tender *crumb*, or texture. Some recipes are formulated for all-purpose flour, but those cakes are more compact and less tender than cakes made with cake flour.

Cake flour is milled to have a smoother texture and fewer gluten proteins than all-purpose flour. It is also bleached to improve the color and the grain of the cake. Because the gluten is missing, it cannot form to toughen the cake. Therefore, cakes made with cake flour can be handled more during mixing and panning than those made with all-purpose flour.

CAKE MIXES. Quantity foodservice organizations frequently use mixes to prepare cakes. The mixes produce a very consistent, high-quality product with less food and labor cost. Most cake mixes for quantity food production require only the addition of water. The additional ingredients to be added to home mixes are already included in quantity mixes.

FOAM CAKES. Foam cakes get their name from the egg foams that form the structure of the batter. These foam cakes must be mixed very carefully. Proper beating of the egg whites and careful blending of the foam with the other ingredients reduces the risk of loss of volume. When correctly prepared and baked, foam cakes spring back to their original shape when touched. These cakes should be turned upside down and cooled while still in the pan to help maintain full volume and shape.

Foam cakes have a medium-fine grain and a slightly rubbery texture. Although the volume of cakes can be increased if the eggs are at room temperature before beating, current regulations prohibit this practice with unpasteurized or fresh eggs.

Angel Food Cake. *Angel food cake* is one type of foam cake. Many quantity foodservices serve this cake to people who are on low-fat, low-cholesterol, low-calorie, or diabetic diets. Angel food cakes are make from egg-white foam, cream of tartar, salt, sugar, vanilla, cake flour, and sometimes almond extract, and they contain no fat.

Angel food cake can be prepared in quantity foodservices using a commercial mix. This one-step mix requires only the addition of water. Most mixes produce cakes with excellent volume, flavor, texture, and tenderness, and their major advantages are convenience and consistency of the final product.

When frozen egg whites are substituted for fresh egg whites in angel food cake recipes, a larger amount of frozen egg whites is needed than the recipe specifies to reach the same volume because a small amount of water and egg yolk is always present in purchased egg whites. The adjustment is usually the addition of 2 ounces of egg white for each pound of fresh whites called for in the recipe. (For a more complete explanation of egg-white foams, see Chapter 8.)

Sponge Cake. Sponge cakes are similar to angel food cakes except the egg foam contains egg yolks that have been combined with sugar. These cakes can be made using the whole-egg or separate-egg method. The whole-egg method uses whole-egg foam while the separate-egg method uses a base of egg-yolk foam in which egg-white foam is folded at the last stage of the mixing. The sugar and egg yolks should be beaten at least 10–15 minutes to ensure a stable foam. This long beating helps to incorporate extra air. Beaten egg whites should be gently folded into the beaten egg yolks to prevent any collapse of the egg-white foam that would reduce the final cake volume. Finally, cake flour is folded very gently with a wire whip attachment into the egg foam with the mixer on low speed.

Like angel food cakes, it is best to cool sponge cakes upside down to retain maximum volume in the final product. When quantity foodservices do not have the space to do this, the cakes can be cooled right side up on wire racks in an enclosed cart. They must be handled gently while they are still warm.

Here is a comparison of the methods for making angel food and sponge cakes.

Angel Food Method	Sponge Method
1a. Using a wire whip, combine the flour with half the sugar, to help the flour mix more evenly with the foam.	1a. Beat the sugar, salt, and eggs with the wire whip at high speed 10–15 minutes until they are light and thick.
1b. Then using the whip attachment, beat the egg whites until they form soft peaks.	1b. Add the liquid ingredients while whipping, either in a steady stream or stirred in, as stated on your recipe.
1c. Gradually beat in the portion of the sugar that was not mixed with the flour. Continue to whip until the egg whites form soft, moist peaks but not until stiff.	1c. Fold in the sifted flour, being careful not to deflate the foam.
1d. Fold in the flour-sugar mixture just until it is thoroughly absorbed.	
2. Place the mixed product in ungreased pans and bake immediately.	2. Immediately pan and bake the batter, to avoid loss of volume.

Chiffon Cake. Chiffon cake is usually more tender than angel food. The egg whites in chiffon cakes are whipped until they are a little firmer than those whipped for angel food, and chiffon cakes contain baking powder so they do not depend on the egg foam for all of their leavening. Chiffon cakes can also be made from mixes. Common flavors are lemon, orange, and chocolate.

SHORTENED CAKES. Shortened cakes have a high fat content. These cakes have a finer crumb, more flavor, and finer grain or texture than foam cakes. Several preparation methods can be used, and the method used will determine the product quality, even with identical ingredients. The conventional method of preparing shortened cakes involves three steps. First, the sugar and fat are creamed. Second, the beaten eggs are added. Finally, the flour and liquid are alternately mixed into the recipe, one-third of each at a time. The modified conventional method differs from the conventional method by how the eggs are incorporated. The egg yolks are added to the creamed mixture, while the egg whites are beaten separately to a foam and folded gently into the batter to incorporate more air, which helps to make a lighter final product.

The sides of the pans for shortened cakes do not need to be greased or coated with flour. Many foodservices lightly grease the bottom of the pans, use a pan-release spray, or use paper pan liners to help remove the cake from the pan. Oven temperature needs to be at least 350°F, to help produce carbon dioxide from the baking powder or baking soda. Unlike foam cakes, shortened cakes are not inverted during cooling. If they are inverted, these cakes might fall out of the pan.

ALTERNATIVE CAKE MIXING METHODS. The muffin method can be used for making cakes. Like muffins, all the dry ingredients are blended together and the fat, in the form of a melted liquid or oil, is added to the other liquid ingredients. This two-stage method saves production time compared to the multistep methods. However, in using this method, tenderness and quality in the final product are sacrificed.

DOUGHS, BATTERS, AND PASTRIES

The single-stage method is the same as the cake mix method. The dry ingredients are combined with shortening and all the water or milk. In the final stage the eggs and any remaining liquid are added to the batter and blended for a specified time. Cakes using this method are coarser.

PROBLEMS WITH CAKES. Here is a list of the causes of common problems with cake products.

Problem	Causes
Tough crumb	Too much egg or flour
	Too little fat, sugar, or liquid
	Overbeating
Sticky crust	Too much sugar
Humped cake	Too hot oven
	Too deep pan
	Too much flour
	Too little milk, sugar, or fat
	Overmanipulation
	Pan filled too full
Dark crust	Too hot oven
	Use of molasses or honey
	Cake too near top of oven
Tunnels	Overmixing
	Excessive gluten development
Cake that falls	Too much fat, sugar, or baking powder
	Oven too cool, or cake removed too soon from oven
	Too little beating
Reduced volume	Too much liquid or fat
	Oven too cool, or inadequate leavening agent
	Cake moved before completely cooked
	Batter too dry

Cookies

Cookies are often served as a dessert. They are classified into three categories: bar, drop, and rolled. The method for mixing is very similar to that used in shortened cakes. The fat and sugar are first creamed together, the beaten eggs are added, and finally the dry and liquid ingredients are added alternately in thirds.

ROLLED COOKIES. Rolled cookies need a stiff dough to be properly handled and cut before being placed on the cookie sheet. Much of this stiffness comes from refrigeration of the dough, rather than from extra flour. Too much flour produces a tough, dry, compact end product. Quantity foodservices might choose this type for holiday cookies. Some organizations use purchased frozen cookie dough. After labor and ingredient costs are considered, this purchased product is not only convenient and standardized in quality, but is similar in cost to cookies prepared from scratch.

DROP COOKIES. Drop cookies have a dough that is fairly stiff so they will maintain their shape on the baking surface. It is easy to make uniform drop cookies by using a dipper. Depending on the baking time and the ratio of fat and sugar to flour, the cookies can be either crisp or soft. Many large quantity foodservices use commercial cookie presses to portion this dough. Chocolate chip and peanut butter are two types of favorite drop cookies.

PASTRIES

Most quantity foodservices use large institutional-size mixers, pastry knife attachments, and pie dough rollers to prepare pastries for use in crusts, shells, and tarts. Smaller operations can prepare them in large batches and freeze them until needed or use the home method for dough preparation. The distinctive pastry texture is caused by blending the fat with the flour before adding liquid. This causes a layering of fat and flour, giving the desirable flaky texture.

Pie Dough

The pie-dough-mixing operation should be watched closely to ensure that the fat is not overblended into the dry ingredients. The fat should be blended very little, leaving it in pea-size pieces that will make a crust with a flaky texture. Overmixing results in a crust with a mealy texture.

Good techniques, accurate standardized recipes, and high-quality ingredients are necessary to produce an acceptably tender and flaky pie crust in quantity foodservices. The fat used should be of a commercial grade with good *shortening power* (the ability to combine with sugars and starches).

When preparing quantity recipes of pie dough, the water should be added very slowly. Too little water results in a dry, unflaky crust. Too much water assists in developing the gluten in the flour, resulting in a tough crust.

Many foodservices purchase frozen pie dough shells or ready-to-use pie crusts. These frozen shells and crusts can be kept for long periods of time and used when either equipment or trained staff is unavailable.

An important factor in production of pies is the prevention of soggy pie crusts. This can be avoided by thinly coating the lower crust with melted fat or flour to create a barrier to the liquid. Some recipes call for a higher oven temperature the first 10–15 minutes to bake the lower crust so it will resist soaking. This may not be possible in quantity food production if staggered baking times are used. In quantity production a partially thickened filling may be used to solve this problem. Other solutions are using a standardized recipe that carefully controls the amount of juice in the filling and baking the pie as soon as it is filled.

Puff Pastry

Puff pastry can be purchased frozen in ready-to-use portions or sheets. These can be used to make tarts, small pies, stuffed turnovers, and Wellingtons. Once thawed, the dough can be handled, rolled, and stretched to make many shapes. In quantity-foodservice operations puff pastries also can be used with meat fillings as a main entree.

DOUGHS, BATTERS, AND PASTRIES

SUMMARY

Doughs, batters, and pastries include a broad array of foods. These products contribute many of the B vitamins, fiber, and calories necessary for good nutrition. Yeast and quick breads are important accompaniments to other foods. Pancakes and waffles are typical foods made from batters. Corn bread, many muffins, and other quick breads are also produced from batters. Cakes, cookies, and pastries are favorite desserts and offer variety to the final course of the meal.

Quantity production methods balance efficiency with high quality. Knowing the function of flour, sugar, fat, leavenings, and eggs allows the baker to understand why different methods are used with various types of baked goods.

Flour is the structural ingredient in all of these items. It contributes gluten to most products. Eggs also provide structure. In other products, like quick breads, cakes, cookies, and pastries, less gluten development is desired. The amount of fat and sugar in the product indicate how tender the food will be.

Pancakes and waffle batters use the chemical leavening agents baking soda or powder. In quantity production, these batters are usually prepared from mixes.

Cakes should be light, sweet, and tender. Foam cakes should have a uniform cell structure and a slightly rubbery texture. Shortened cakes are more tender due to the larger amount of fat and sugar.

Cookies are classified into three categories—bar, drop, and rolled. The mixing method for cookies is similar to that for shortened cake, that is, creaming fat and sugar and then adding dry and liquid ingredients. Cookie mixes or frozen dough are frequently used in quantity production. Dry mixes compare favorably in ingredient and labor costs, however, the flavor and variety may be more limited.

Pastries are used in many rich, high-calorie desserts. Preparing high-quality pastry is more challenging than preparing optimum cakes and cookies because the mixing must be very carefully controlled and ingredient temperature also influences quality. In pastry, the flour is combined with the fat to produce layers of these ingredients necessary for the characteristic flaky texture.

LEARNING ACTIVITIES

Activity 1: Comparing Muffins Made from a Mix with Muffins Made from Scratch

1. Prepare two batches of blueberry muffins.

    A. Prepare one batch using a blueberry muffin mix, following the directions on the package.

    B. Prepare another batch of blueberry muffins, using the recipe.

2. Record your comparison of the two kinds of muffins on the charts.

Activity 1, Step 2

Recipe Type	Yield	Cost per Muffin, Food and Labor	Preparation Time
Scratch			
Muffin Mix			

Activity 1, Step 2

Recipe Type	Texture	Appearance	Flavor
Scratch			
Muffin Mix			

Blueberry Muffins

Yield: 10–11 doz
Portion size: #24 dipper or 1½ oz
Baking temperature: 400 F
Baking time: 20–25 min

Amount	Ingredients	Amount	Procedure
....	Flour, all-purpose	3 lb 8 oz	1. Combine flour, sugar, baking powder, salt, and nonfat dry milk in 12-qt mixer.
....	Sugar, granulated	1 lb 4 oz	2. Blend on low speed, using flat beater.
....	Baking powder	4 oz	
....	Salt	1¼ oz	
....	Milk, nonfat, dry	6½ oz	
....	Eggs	11 oz	3. Beat eggs until light and foamy.
....	Water, room temperature	1 qt 2½ c	4. Add water to eggs; mix thoroughly. 5. Add combined eggs and water to dry ingredients.
....	Shortening, hydrogenated, melted	1 lb	6. Add melted shortening which has been cooled to room temperature. 7. Mix on medium speed for 10 sec. *Note:* Mixture will not be smooth.
....	Blueberries, frozen	1 lb 8 oz	8. Add blueberries (do not defrost) and fold into batter by hand. Mix only until distributed in batter.
	Total Weight	12 lb	9. Portion into greased muffin tins using #24 dipper. 10. Bake at 400 F for 20–25 min or until golden brown.

Source: *Standardized Quantity Recipe File* (1971).

3. Set up a taste panel, as described in Chapter 1 on recipe standardization. After the panelists have tasted the two types of muffins, complete the charts. Did the panelists prefer one over the other?

DOUGHS, BATTERS, AND PASTRIES

Activity 1, Step 3

Blueberry Muffins from Mix			
Panelist	Texture	Appearance	Flavor

Activity 1, Step 3

Blueberry Muffins from Scratch			
Panelist	Texture	Appearance	Flavor

4. Wrap at least 6 mix and 6 from-scratch muffins and freeze them for 3 or more days. Then compare them for texture, appearance, and flavor again. Which of the two preparation methods has the better *holding* (keeping) quality? _____

5. Which had the lowest food and labor cost per muffin? _____

6. What are three examples of situations when a muffin mix would be used in a quantity foodservice?

    1. _____

    2. _____

    3. _____

7. If you had blueberry muffins on the menu in your facility, should mix or from-scratch be used? Why?

    _____

    _____

Activity 2: Comparing Yeast Dough and Quick-Bread Batter

1. Fill in the chart, listing the major differences between yeast doughs and quick-bread batters.

Activity 2, Step 1

	Yeast Bread	Quick Bread
Amount of liquid		
Amount of fat		
Amount of sugar		
Type of leavening		
Method of mixing		
Amount of gluten development		

2. What does gluten do in yeast breads?

3. What does gluten do in quick breads?

4. List three leavenings used in bakery products.

5. How do each of these three leavenings cause the product to rise?

Activity 3: Comparing Cake and Cookie Ingredients

1. Fill in the chart, using the recipes for Devil's Food Cake and Chocolate Chip Cookies.

DOUGHS, BATTERS, AND PASTRIES

Activity 3, Step 1

Ingredient	In Cakes	In Cookies
Sugar		
Fat		
Flour		
Leavening agent or agents		

2. If both batters were overmixed, would the cake or the cookie quality be more harmed? Why?

___

___

___

Devil's Food Cake

Yield: 108 portions *or* 3 12×20×2" pans
Portion size: 36 per pan
Baking temperature: 350 F
Baking time: 40–50 min

Amount	Ingredients	Amount	Procedure
....	Flour, cake	2 lb 6½ oz	1. Combine flour, soda, sugar, and cocoa in 20-qt mixer (with extender on bowl).
....	Soda, baking	2 TBSP 2½ tsp	2. Blend on low speed using flat beater.
....	Sugar, granulated	4 lb 3 oz	
....	Cocoa	14½ oz	
....	Shortening, hydrogenated, room-temperature	14½ oz	3. Add shortening and vanilla to flour-sugar-cocoa mixture.
....	Vanilla, pure extract	3 TBSP 2 tsp	4. Mix on low speed. *Approx time:* 2 min
....	Buttermilk	2⅓ c	5. Add buttermilk and mix on low speed until blended.
			6. Beat on high speed, then scrape sides. *Time:* 2 min
....	Water, lukewarm	2⅓ c	7. Combine water, unbeaten eggs, and buttermilk.
....	Eggs	2 lb 3 oz	8. Add slowly to batter while mixing on low speed; scrape sides.
....	Buttermilk	2⅓ c	9. Beat on high speed. *Time:* 3 min
	Total weight	14 lb 4 oz	10. Scale into greased pans. *Amt per pan:* 4 lb 12 oz
			11. Bake at 350 F for 40–50 min.
			12. Cool cake.
			13. Frost as desired.

Source: *Standardized Quantity Recipe File* (1971).

Chocolate Chip Cookies				
Yield: 100 cookies with #40 dipper *or* 70 cookies with #30 dipper Portion size: ¾ oz with #40 dipper *or* 1–1¼ oz with #30 dipper				Baking temperature: 400 F Baking time: 10 min
Amount	Ingredients		Amount	Procedure
....	Shortening, hydrogenated Vanilla, pure extract		13 oz 2 tsp	1. Have ingredients at room temperature. 2. Place shortening and vanilla in 5-qt mixer. 3. Cream, using flat beater, on medium speed.
....	Sugar, granulated Sugar, brown		12 oz 9 oz	4. Add sugars gradually, beating on high speed until light and fluffy.
....	Eggs		8 oz	5. Add unbeaten eggs and continue beating on high speed until light.
....	Flour, all-purpose Soda, baking Salt		1 lb 4 oz 2 tsp 2 tsp	6. Combine flour, soda, and salt in a bowl; blend.
				7. Add combined dry ingredients to creamed mixture on low speed. 8. Beat on medium speed until well blended.
....	Nuts, coarsely chopped Chocolate chips		4 oz 12 oz	9. Add nuts and chocolate chips; mix on low speed until uniformly distributed.
	Total weight		5 lb	10. Portion onto greased sheet pans (18×26×1"), leaving approx 2" between cookies. *No. per pan:* 28 cookies 11. Bake at 400 F for 10 min.

Source: *Standardized Quantity Recipe File* (1971).

REVIEW QUESTIONS

Multiple Choice

1. Yeast breads are baked dough normally made from flour and

    A. Water, yeast, and baking powder
    B. Water, sugar, and yeast
    C. Milk, salt, and yeast
    D. Water, yeast, and baking soda

DOUGHS, BATTERS, AND PASTRIES

2. Why does whole wheat flour have a shorter shelf life than white all-purpose flour?

   A. The inferior grade of gluten deteriorates quickly.
   B. The endosperm becomes rancid.
   C. The additional outer layer of bran becomes too coarse.
   D. The wheat germ becomes rancid.

3. Which step is a feature of the muffin method of mixing quick breads?

   A. Fat is cut into the dry ingredients before the addition of the moist ingredients.
   B. Oil and sugar are creamed before the addition of the dry and moist ingredients.
   C. Dry ingredients are mixed together, and then the eggs, milk, and oil are added all at once.
   D. Creamed fat is cut into the dry ingredients before the addition of the moist ingredients.

4. Which of the following would most likely be responsible for making a cake fall?

   A. Too much flour
   B. Excessive salt
   C. Excessive sugar
   D. Extra gluten development during mixing

5. Cream of tartar is added to angel food cake to

   A. React with the baking powder
   B. Produce larger air cells in the egg-white mixture
   C. Stabilize the egg whites and increase the volume
   D. Reduce the acidity of the mixture to add stability

6. Which of the following ingredients usually accounts for the leavening in waffle batter?

   A. Baking soda and water
   B. Baking powder and liquids
   C. Cake flour and baking soda
   D. Brewer's yeast and sugar

7. Which flour is most appropriate for making apple turnover dough?

   A. Whole wheat
   B. Cake
   C. Pastry
   D. All-purpose

8. Which of the following does *not* create the flaky texture of pastry?

   A. Layers of fat and flour
   B. Coating the gluten with fat
   C. Carbon dioxide leavening
   D. Combining the hot liquid with the fat

10. BEVERAGES AND CONVENIENCE FOODS

BEVERAGES

All quantity foodservice operations offer a number of beverages to their clients. Beverages vary in popularity from one region of the country to another. For example, sweetened iced tea is usually served in the South and unsweetened iced tea in the North. The most popular beverages are coffee, sodas, hot and iced tea, milk, hot chocolate, and fruit juices.

Coffee

Regular and decaffeinated coffee should be available in all foodservice facilities. Both coffees are a roasted blend of several different coffee beans selected to create a desired color, aroma, and flavor. The coffee beans are usually shipped to the United States while still raw. They become dark brown after roasting, as a result of the browning of the sugar in the beans and the breaking down of the starch. This process is called *dextrinization*. The aroma and flavor are enhanced by grinding the roasted coffee beans to expose the maximum amount of surface area for brewing. The fineness of the grind must match the requirements of the coffee equipment.

A percolator uses a coarse grind of coffee that has a limited amount of exposed surface of the coffee beans. A fine grind, which increases the exposure to the grounds, is used in drip and pour-over coffee machines. These machines dispense hot water so it drips through the grounds into the pot, exposing the grounds to the hot water only once. These units usually require 4–6 minutes for brewing a 1- to 2-quart pot of coffee, the usual type in quantity-foodservice coffee machines.

Using the pour-over method, 1 pound of coffee grounds usually makes 2 gallons of coffee. The correct water temperature for brewing coffee is 195–200 F. Once brewed, coffee should be held at 185–190 F. Hotter temperatures cause the coffee to scorch and the natural oils to separate and float to the top. To maintain desirable taste, aroma, and flavor, brewed coffee should not be held longer than 1 hour.

Quantity foodservices often use automatic commercial coffee machines that only require placing a premeasured packet of coffee grounds in a disposable filter and turning on the machine. The less an individual needs to do in measuring coffee and timing the process, the better the chances for a standardized cup of coffee. Premeasured coffee packets, premeasured water, and automatic timers reduce the risk for error.

One of the most critical elements in brewing coffee or any beverage is that the machine and containers must be clean. Coffee machines are very susceptible to buildup of acids and oils that turn rancid. A regular cleaning schedule will prevent the development of these rancid and bitter flavors, which can ruin a freshly brewed beverage.

Coffee pots and coffee machines are cleaned by using a commercial cleaning packet provided by the coffee distributor or coffee machine manufacturer or by using soapy water to dissolve the oils that cling to pots and machines. It is important to rinse all equipment very thoroughly to eliminate the possibility of soapy residue. Glass coffeepots can be cleaned easily by swirling crushed ice and a mild soap solution in each pot and rinsing with clear water. Cleaners should be stored in a separate area away from all foods and beverages.

The manufacturer's instructions should be followed to remove lime deposits in coffee machines. Water lines should be filtered to help reduce the buildup of lime, calcium, and other minerals.

Coffee machines should not be connected to hot water unless the machine has a built-in filter for hot water. Many hot water systems are softened with salt, which will cause off-flavors in coffee.

Coffee is available in a number of instantly soluble forms, such as flakes, crystals, powders, and frozen liquids. There are especially designed coffee machines to reconstitute these forms of instant coffee. These machines are smaller, more compact, and less expensive than the larger ones required for service with regular coffee grounds. Many of these machines also require less cleaning and maintenance and produce a more standardized cup of coffee. Even though these machines are less expensive, the cost per cup of coffee may be more expensive because instant coffee is more expensive than ground. In addition, because of the loss of many of the flavoring substances during the processing of these coffee concentrates, many people do not like their flavor. It is wise to evaluate each system for coffee quality and for labor, product, and equipment costs.

Decaffeinated coffee is another form of coffee all foodservices should offer. Some clients may need a low-caffeine or a no-caffeine beverage. Today many people select decaffeinated coffee to eliminate the stimulating effect of caffeine. Decaffeinated coffee, like regular coffee, is available as grounds, flakes, crystals, powders, and liquids. Most foodservices choose to carry only one type of decaffeinated coffee for economies of space and inventory. Many methods of removing the caffeine from the coffee bean are used. Because of the possible toxicity of the older petroleum method, a water-extraction method is preferred but the product is more expensive.

One of the best ways to judge the types of coffees on the market is to have a taste panel sample the various brands without knowing what brand they are tasting. Coffee quality and taste should be considered, along with product price. Because the color and strength desired in coffee will vary from one individual to another, it is best to develop an agreed-upon standard for a foodservice and to provide the written description of what is considered acceptable coffee prior to the beginning of the test. Coffee vendors who want to provide coffee for the foodservice should *not* be allowed to conduct the test because they might attempt to influence the results in their favor. They should, however, provide samples for comparison.

Here are guidelines for making good coffee.

- Use clean, fresh water that is not hard and has not been softened with salt additives.
- Use clean coffee machines, coffee pots, and china.
- Use only fresh coffee grounds of the appropriate grind for the brewing method.
- Use the appropriate roast of coffee bean. A dark roast will produce a stronger coffee with a dark brown color.

- Use premeasured packets or the amount of grounds recommended by the coffee distributor or coffee machine manufacturer.
- Brew coffee at the proper temperature, 195–200 F.
- Hold coffee at a temperature of 185–190 F and never allow it to boil.

Tea

Many different teas are available. The three most common types are oolong, black, and green. The same variety of tea plant, with different processing, is used to make all three types. Although specialty and low-caffeine teas have gained popularity in recent years, black tea is still the most popular in the United States.

Teas are usually a blend of two or more types of tea leaves. And, like coffee, a good tea begins with a high-quality, fresh product. Unlike coffee, tea does not contain oils, making it much easier to clean and maintain the brewing and serving equipment.

To have clean, clear water for brewing tea and coffee, quantity foodservices may need to use a water filter. As in making coffee, the best water is soft but not softened. Hard water will react with the tannins in the tea or coffee, producing a cloudy product with a film on the surface.

Most tea manufacturers recommend bringing the water to a boil and immediately pouring it over the measured tea or tea bags. The tea should be steeped at 185–198 F for 3–5 minutes. Longer steeping periods produce a stronger flavor.

Serving hot, near-boiling water to clients along with individual tea bags allows them to brew tea to their own preferences. Many people brew a second cup from one tea bag.

Iced tea is another beverage offered by most quantity foodservices. It can be brewed fresh using commercial-size tea bags, or it can be made using instant tea crystals or concentrated liquids. Many operators use instant tea and commercial dispensers directly on the beverage serving line. This allows clients to obtain their tea freshly prepared, adding the preferred amount of ice.

Milk Beverages

The milk types most commonly offered as beverages are whole, 2%, and skim milk and low-fat chocolate milk. Because whole milk has an approximate fat content of 3.5%, many people who are controlling their fat and cholesterol intake choose either skim or one of the low-fat milks. To be labeled as skim milk, the product must contain less than 1% fat. Low-fat milk is usually at the 1–2% level. Low-fat chocolate milk is usually made with 2% milk. This chocolate milk is generally lower in sugar than chocolate-flavored whole milk.

In quantity foodservices, milk products are either offered in individual cartons or served in bulk dispensers using five-gallon plastic bag containers. Care must be taken when installing the bag containers into bulk dispensing units to prevent the bags from breaking. The dispensers also need daily cleaning to maintain proper sanitation standards. Local health department officials can be helpful in providing sanitation methods for proper milk dispensing.

Careful checking of expiration dates is an important part in serving milk products. No milk product should be served beyond the expiration date, even though the product may seem to have an acceptable flavor and appearance. It is a good idea to make the rotation of the milk products a responsibility of the milk supplier. However, the foodservice manager also must check the dates to ensure that milk products are being rotated properly.

HOT CHOCOLATE. In quantity foodservices, hot chocolate is usually served from liquid concentrate or powdered forms. Dispensers are available for both products, but powdered hot chocolate mixes are more commonly packaged in individual servings. Bulk dispensers, however, decrease chances for theft of the small individual envelopes.

The preparation of hot chocolate is designed to prevent the hot chocolate mix from settling to the bottom of the cup. Starches in the mix are pregelatinized and expand with the addition of moisture, to help in prevent settling. Hot chocolate made from fresh milk should not be held longer than 1 hour before it is consumed. This prevents scum from forming on the surface, the result of milk proteins separating.

Carbonated Beverages

Regular and low-calorie carbonated beverages are a daily part of many people's diets. These beverages often are served from remote syrup and carbonated gas dispensers in the cafeteria or dining room. To offer a wide variety, most quantity foodservices select dispensers that provide four or five different choices. Usually at least one choice is a low-calorie diet drink.

The *Brix* (the syrup/water ratio in the drink) should be checked on a weekly basis to ensure that the proper syrup concentration is maintained. This helps to control product costs and also ensure the quality of these beverages.

A major problem with bulk dispensers is that beverages are usually served unsealed. These cups and serving vessels will not retain the carbonated gas in the beverage for extended periods of time. Therefore, quantity foodservices that serve remote or satellite locations usually provide these beverages in sealed cans or plastic bottles.

Cans also are used for tray service. The smaller 8- or 10-ounce plastic bottles or aluminum cans are available from some manufacturers. This smaller size is often preferred in nursing facilities.

CONVENIENCE FOODS

Types of Convenience Foods

Convenience foods are available in many forms. Any food item that has been partially processed before it is delivered to the foodservice is a convenience food. These foods are purchased either as total convenience foods that are ready-to-serve or as partially prepared ingredients for use in other recipes. Canned puddings, pie fillings, and sauces; pregelatinized flour; dehydrated soups and potatoes; and frozen eggs, desserts, rolls and dough, and entrees are examples of the many convenience foods that are available.

Convenience products for baking are especially helpful in providing a large variety of freshly baked products for the menu. Frozen croissants, doughs, and puff pastry are a few of the choices that might otherwise not be possible to serve. Puff pastry, for example, may be filled with a meat mixture, folded, sealed, and baked for service as a main entree or may be used in desserts, such as apple turnovers.

Partially processed fruits and vegetables prepared by a produce company also qualify as a convenience food. Products such as carrot salad, potato salad, coleslaw, and macaroni salad are other examples of the convenient produce foods available. Some produce companies will make products according to individual specifications. This can help the foodservice manager with a limited staff to satisfy the production demands of an extensive menu.

Storing and Preparing Frozen Convenience Foods

Frozen food should be stored at 0 F or lower; dry and canned products in a dry, cool area; and refrigerated foods below 40 F.

Frozen convenience foods should be placed in the freezer immediately after being received. To avoid texture changes, these foods should not be allowed to partially or completely thaw and then be refrozen. Once a frozen convenience food has been thawed, it should be used within 72 hours. The threat of spoilage from microorganisms is greatly increased in a thawed product.

Frozen convenience foods have a better quality and taste if they are *tempered,* or allowed to defrost slowly, while under refrigeration of 28–30 F. This minimizes the cost and time of reheating while maintaining necessary sanitary standards. The amount of time allowed for defrosting while under refrigeration depends on the size and density of the frozen food. For example, a spinach soufflé will take considerably less time to defrost than a case of frozen, precooked turkey breasts. The spinach may take only 24 hours to defrost while the turkey may take as much as 48 hours. Most convenience products provide precise defrosting instructions on the container.

Convenience foods that are already partially cooked can spoil rapidly. They should not be left outside their storage areas for more than 1 hour. *Reconstituting,* which is the process of preparing the food for service, should be monitored carefully to ensure that optimum standards are maintained. As with any food item, this stock should be dated when received and rotated on a first-in, first-out basis.

Not all convenience foods are purchased. If quantity foodservices have qualified personnel available and the necessary freezer space, foods such as lasagna can be cooked and frozen for later service. Close monitoring is necessary for proper sanitation and preparation procedures, correct storage temperatures, and regular stock rotation.

Many quantity foodservices use canned convenience products to supplement their menu. These products are very stable and have a long shelf life if stored in a clean, dry, and cool storage area. Canned products, such as spaghetti sauce, ravioli, clam sauce, and chili, can be heated quickly to provide instant relief for production shortages during peak work periods or staff shortages.

Dried, freeze-dried, and dehydrated convenience foods, such as onion flakes, dry soups, instant potatoes, and stuffing mixes, have a very long shelf life if kept in a clean, dry, and cool storage area. The manufacturer's instructions for rehydrating should be followed. It is important to remember that if a quantity recipe calls for a certain amount or weight of *fresh* product, this would be equal to the same weight of *rehydrated* product, not the dry weight.

Many manufacturers have designed their convenience foods to be prepared in a microwave. Some specify that the product is first heated using the defrosting cycle and then cooked on the regular cooking cycle.

When to Use Convenience Foods

Both the quality and the price of convenience foods must be considered before they are introduced on the menu. If skills of foodservice workers are very limited, the manager might choose to place a number of convenience foods on the cycle menu to offer a variety of selections. Many operators offer one or two entrees prepared from scratch, in combination with convenience foods. However, the quality and acceptability of these convenience foods must be watched closely. Some clients and administrators believe convenience products should not be substituted for those produced by the facility.

A careful analysis is necessary to compare the costs of convenience and conventional products, using standardized portions. Manufacturers specializing in frozen doughs and puff pastries, for example, can produce a high-quality product for a reasonable cost. However, many foodservices do not have personnel properly trained to handle these delicate products, the correct equipment and ingredients, or the necessary time. The labor and ingredient costs and quality must be compared with from-scratch products when deciding whether to use convenience bakery items. Any analysis should also take into account the color, texture, flavor, eye appeal, ease of portioning, and mouth feel of the products.

It is unlikely that convenience foods will replace all conventional cooking methods. These foods will, however, be used more and more to supplement the menu, providing variety and appeal.

BEVERAGES AND CONVENIENCE FOODS 167

Quantity foodservices should continually evaluate the advantages and disadvantages of using convenience foods.

SUMMARY

Serving beverages of best quality is possible only when dispensing machines are properly cleaned and maintained. Machines for coffee and carbonated beverages help produce and serve beverages made correctly and at the best temperature. Milk can be served from individual cartons or 5-gallon dispensing machines. Careful rotation of stock to avoid having milk on hand past its expiration date is important.

Convenience foods in a wide range of preparation stages are used in quantity foodservice. Optimum convenience products can be served when high-quality items are purchased, stored correctly, and heated appropriately.

LEARNING ACTIVITIES

Activity 1: Comparing Brewed Coffee with Instant Coffee

1. Prepare one pot of coffee in your coffee machine, following established coffee brewing procedures. (Note: You can use a large coffee pour-over urn. Be certain, however, to keep accurate records on the total volume of coffee produced for cost calculations.)

    A. Weigh the dry coffee grounds before placing them in the coffee filter.

    B. Measure in fluid ounces the exact volume of coffee produced, for later cost calculations.

2. Make the same amount of coffee from freeze-dried coffee, following the directions on the jar or package.

3. If frozen coffee concentrate is available, make a third pot of the same volume.

4. Fill in the chart to compare the coffees. Determine the cost of the pots of freeze-dried and frozen concentrate coffees, using costs and yields provided by the manufacturer or supplier.

5. Repeat the test for color, clarity, aroma, taste, and aftertaste after holding the three coffees hot for 30 minutes.

6. Which coffee required less preparation time and cleanup time? _____

7. Which coffee had the best flavor? _____

8. What differences (if any) did you notice in the aromas?

_____

_____

Activity 1, Step 4

	Regular Grounds	Frozen	Freeze-dried
Volume, in fluid ounces			
Cost per ounce of brewed coffee			
Color			
Clarity (clearness)			
Aroma			
Taste			
Any aftertaste			

9. Why should quantity foodservices maintain an alternative backup coffee?

_____

_____

Activity 2: Comparing a From-Scratch and a Frozen Convenience Meat Loaf

1. Prepare the recipe for meat loaf in Chapter 7, Activity 2. Make one 12×20×2½" pan of 48 portions.

2. Compare the from-scratch recipe with a frozen convenience meat loaf, in the same-size pan, if possible. If the convenience meat loaf is a different size, note the weight of each product, excluding the pan weight.

   A. Portion the from-scratch meat loaf according to the recipe. Weigh one portion to compare for comparison with the weight of one portion of the convenience meat loaf.

   B. Keep accurate records of the labor used in the preparation of each product and calculate labor costs according to the hourly wage of the cooking staff. Use the following formula to calculate the labor cost per minute:

   Hourly wage/60 minutes = Cost per minute

   Then multiply by the preparation minutes to calculate the total labor cost.

   C. Complete the chart comparing the two meat loaf products.

BEVERAGES AND CONVENIENCE FOODS 169

Activity 2, Step 2C

	Convenience	From Scratch
Total food cost		
Total labor cost		
Servings per pan		
Weight per serving, in ounces		
Total cost per ounce, food and labor		

    D. Compare the differences in the quality of the two products using the chart.

Activity 2, Step 2D

	Convenience	From Scratch
Color		
Appearance		
Ease of cutting		
Ease of serving		
Retention of shape on plate		
Aroma		
Flavor		

3. Which of the two products is least expensive, from-scratch or convenience?

4. What other factors might affect the overall cost of the frozen convenience meat loaf?

5. What are some of the reasons convenience foods are used to supplement from-scratch products?

6. Which of the two products did you consider to have the highest quality, from-scratch or convenience?

7. List five convenience foods you might use to supplement your menu.

REVIEW QUESTIONS

Multiple Choice

1. What is the correct water temperature for brewing coffee?

   A. 212 F
   B. 185–200 F
   C. 195–200 F
   D. 200–210 F

2. Why do many quantity foodservice managers choose to use liquid coffee concentrate instead of conventional ground coffee?

   A. It can be stored for longer periods of time before brewing.
   B. One pound of ground coffee is less expensive than 1 pound of liquid concentrate.
   C. The aroma is better than in coffee made from grounds.
   D. It produces less waste than ground coffee.

3. What is the most popular type of tea in the United States?

   A. Ginseng
   B. Chinese
   C. Green
   D. Black

4. In brewing iced tea, which procedure should be followed?

   A. Pour cold water into the hot tea concentrate.
   B. Pour hot tea concentrate into the cold water.
   C. Pour hot tea concentrate over ice.
   D. Pour cold water and hot tea concentrate into a glass or stainless steel container and refrigerate for 24 hours.

5. Whole milk contains what percent butterfat?

   A. 1.5–2.5%
   B. 2.5–3.5%
   C. 3.5–4.0%
   D. 4.0–4.5%

6. Checking the Brix scale in carbonated beverages refers to the amount of

   A. Carbon dioxide gas in the storage tanks
   B. Syrup in the syrup/water mixture
   C. Carbon dioxide in the syrup/water mixture
   D. Water in the syrup storage canister

7. When convenience food directions specify thawing before cooking, which of the following defrosting methods is best?

   A. In a microwave
   B. At room temperature
   C. In cool water
   D. In the refrigerator

8. When doing the daily inventory of the walk-in refrigerator, you notice a partial pan of cooked lasagna, a convenience product, dated 2 days ago. What should you do with this food?

   A. Serve it at a meal that same day.
   B. Throw it away as soon as possible.
   C. Keep it refrigerated and serve within 2 days.
   D. Keep it refrigerated and serve within 3 days.

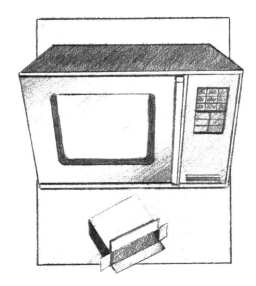

11. MICROWAVE COOKING

Microwave cooking has been used in foodservice organizations since the early 1970s. The microwave oven is known for its speed and ability to make very high quality cooked vegetables. It is the preferred method of reheating small quantities of food. In quantity food production, microwave ovens are useful for cooking and for reheating individual portions of food.

HOW MICROWAVE OVENS WORK

Microwaves are very short, high-frequency radio waves. The microwave oven is similar to a small broadcasting system for these radio waves. This is the same type of energy used to broadcast AM or FM radio. The main differences are that microwaves are shorter and are broadcast and received inside the oven.

The microwaves are transmitted when the oven door is closed and the oven is turned on. The transmitter, called a *magnetron,* sends a signal to the receiver inside the oven. The receiver acts like a reflector, bouncing the waves back into the oven. The waves bounce off the special metal walls inside the oven until they come in contact with other metals. Metal will stop the microwaves and reflect them back toward the receiver. Microwaves will pass right through paper, glass, and plastic because they contain no moisture. When microwaves pass through an item, it does not get hot, but when microwaves bounce off metal placed in the oven, it gets very hot even though the special type of metal walls of the oven stay cool. Glass, paper, and plastic dishes used for cooking in a microwave will not be heated by the microwave but may get hot from coming in contact with the hot food.

When the microwaves come in contact with moisture, fat, and sugar, they vibrate the molecules of moisture very quickly, about 2½ billion times a second. The fast movement causes friction between the molecules and this friction makes the molecules get hot. The heat then cooks the food.

Microwaves will reach only molecules near the surface of the food, only about the outside inch. The waves penetrate the food from the top, bottom, and all sides because the waves bounce around in the oven.

The waves only heat about the outside inch, and the rest of the food is cooked by conduction of heat throughout the food. *Conduction* is the same type of cooking process that occurs when a range top is used. It means the food is heated by coming in contact with something hot. On the range, the food gets hot from touching the hot pan. In a microwave, the inside of the food gets hot by being next to the part the microwaves heated.

Figure 11.1 shows the effects of direct heat (range top cooking), baking in a conventional oven, and microwave cooking.

Figure 11.1. Heating patterns of a range top, a conventional oven, and a microwave oven.

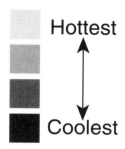

Range top: The product heats by conduction, starting at the bottom. The food interior is heated by conduction, causing the top to be cooked last.

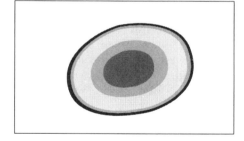

Conventional oven: The food is heated on the outside by the hot, dry oven air. The inside of the food heats from the outside in, by conduction. The center is cooked last.

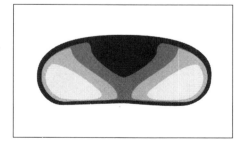

Microwave oven: The first food to heat is ¾–1" inside the surface. Heat is conducted toward the surface and the center. The areas of the food of highest moisture, sugar, or fat heat first.

Oven Wattage

The amount of power output in microwave ovens is called *wattage*. The higher-wattage ovens cook faster. Ovens range from 450 to 1500 watts. Ovens with less than 650 watts are best suited for reheating snacks, sandwiches, and beverages. Cooking other foods in these low-wattage ovens may save little time, compared with other cooking methods. Ovens with 650–750 watts are intended for home use in cooking and reheating foods. Ovens manufactured for quantity-foodservice organizations have 900–1500 watts. The high power of these ovens allows 6–10 portions of food to be cooked or heated in the oven, yet still save time. Single portions of food also cook or reheat very quickly in these high-voltage ovens.

Most microwave ovens can be set to use less than full power. Usually power levels available are percentages of 100% (full) power. The percentage means the amount of time each minute the oven is producing microwaves. The remaining time the oven is not cooking. It is like turning the oven on for a few seconds and then turning it off so the heat in the food can be conducted throughout the food before the outside is overcooked. For example, 90% power means the oven is on 90% of each minute and off 10% of each minute. The lower levels of power are used when food needs to be cooked slowly. Low-power levels are used to defrost foods. This prevents the outside from cooking before the inside is thawed. For thawing, 10–30% power is the best choice. At 30% power the oven is on for 3 seconds, then off for 7 seconds. This cycle repeats until the total time is reached.

Microwave cooking and reheating directions usually specify only one or two wattage levels. To adjust the cooking time for an oven of a different wattage, a conversion factor is used.

CONVERTING MICROWAVE RECIPES FOR VARIOUS WATTAGE OVENS. To identify the correct microwaving time when the oven wattage is different from the wattage specified in a recipe is a two-step process. First the wattage of the oven that will be used is identified. The wattage is usually listed in the manual or guide for the oven. The wattage also can be obtained by asking the manufacturer or place of purchase.

If the wattage cannot be obtained from these sources, it can be identified by timing how long it takes to boil 1 cup of ice water in the oven. Because the actual wattage output of the oven is affected by the electric current supplied to the oven, this procedure also is recommended when the oven shares the electric circuit with other appliances. To determine the oven wattage, 1 cup of 33–40 F water in a glass or microwave-safe bowl is placed in the oven. The heating of the water is carefully timed. The uncovered water is heated using the 100% power level until it comes to a full rolling boil. The minutes and seconds it takes to boil is compared with the information in Table 11.1. Many microwave cookbooks contain similar tables.

To identify the wattage of the microwave oven, find the time in the left column that is closest

Table 11.1. Time needed to boil 1 cup of ice water at several wattage levels: Conversion factors for 700-watt and 575-watt recipes

Heating Time	Estimated Oven Wattage	Adjustment Factor for Recipes Specifying	
		625- to 700- Watt oven	475- to 575- Watt oven
1 min to 2 min	1300–1500	0.500	0.320
2 min to 2 min, 30 sec	1000–1200	0.637	0.408
2 min, 30 sec to 3 min	875–925	0.781	0.500
3 min to 3 min, 30 sec	725–800	0.877	0.561
3 min, 30 sec to 4 min	625–700	1.000	0.640
4 min to 5 min	475–575	1.563	1.000
5 min to 6 min	425–475	1.639	1.049

to the time required to boil the ice water. The wattage listed in the same row is the approximate wattage of the oven used to heat the water. In the two other columns are the factors used to convert the cooking time in a recipe to the correct time for the oven to be used. Use the conversion factor from the far-right column when the recipe specifies a 625- to 700-watt oven. The conversion factors in the far-right column are used when the recipe is written for a 475- to 575-watt oven.

For example, if it took between 3 minutes and 3 minutes 30 seconds to boil ice water in a nursing facility's microwave oven, the oven must be approximately 725–800 watts. To convert the cooking time in a recipe written for a 700-watt oven, multiply the recipe cooking time by the conversion factor 0.877 to determine the cooking time for the 725–800 watt oven. If the cooking time in the recipe is 5 minutes, the cooking time in the facility's oven is 5 minutes times 0.877, or 4.385 minutes, or 4 minutes 23 seconds. (To convert .385 minutes to seconds, multiply it times 60 seconds; .385 minutes × 60 = 23 seconds.)

Microwave Cooking Principles

Microwave cooking involves many of the same principles as other methods of cooking. Because the cooking is so fast, understanding these principles is even more important than in traditional cooking.

ARRANGING. When several items are microwaved at the same time, their arrangement affects cooling time. Arranging food in a circle is recommended. If heating several foods on a plate, place the dense foods or large pieces nearest the rim.

BURSTING. The buildup of steam inside foods may cause the surface to break open. Because the surface may cook much faster than the center of the food, it cannot expand as steam builds up inside. Bursting happens in foods like whole apples, whole eggs, egg yolks, and potatoes. It also can happen when foods are covered too tightly or are in a closed plastic bag or jar. To prevent bursting, the surface or cover should be pierced with a sharp knife or fork, or removed before cooking, or the food should be partially uncovered.

DENSITY. A food's density describes whether a food is solid or is light and airy. Less dense, light, and airy foods, (also called *porous*) such as bread and cake, absorb microwaves easily and cook quickly. Foods that are denser, such as potatoes, casseroles, and meat, cook more slowly. The microwaves are absorbed by the food because there are fewer air spaces for the microwaves to pass through.

HEIGHT. The microwaves enter from the back of the oven near the top. Food nearest the source of the waves usually cooks fastest. Foods that are tall enough to nearly fill the oven will need to be turned over to be cooked more evenly.

MOISTURE. Foods with high moisture levels microwave very well. The microwaves are attracted to the moisture. Microwaves also seek out food with high sugar and high fat levels. Dry foods, such as cooked roast turkey, heat better if a moist sauce or gravy is added. Dry foods also cook more evenly if they are covered.

QUANTITY. All of the microwave watts are sent into the oven every time it is turned on. When one small portion of food is being cooked, all the waves are concentrated on that small item. If 6 portions are cooked, the same number of watts must be shared by all 6. Six items will take longer to cook than one item, but not 6 times longer. In large quantity, microwave cooking may even be slower than with traditional methods.

SHAPE. Foods that are round or circular microwave the most evenly. Ring-shaped containers, like ring molds and bundt pans, help dense foods cook more evenly. Square or rectangular pans allow the food in the corners to cook before the center is done. Unevenly shaped foods like roasts, chicken breasts, and broccoli spears result in uneven cooking. These foods require turning over and rearranging during the cooking time.

SHIELDING. The smaller parts of large or irregularly shaped foods should be covered to avoid overcooking them. The most frequent need for this shielding is when square pans are used. Shielded corners will not overcook before the center is done. Foil is usually used for shielding. It is important to follow the instructions for use of foil in the oven.

SIZE. Small pieces of food cook faster than large pieces of the same food. When several pieces of food are cooked at the same time, they should be the same size. Using dippers to portion meatballs, muffins, or other foods helps make them all the same size. When more than one potato is baked in a microwave oven, they should be of similar size and uniform shape.

STANDING TIME. Allowing foods to stand after microwaving is an important part of the cooking method. Foods continue to cook after being taken out of the oven. The standing time also allows the temperature to become uniform, in case hot-spotting occurred.

TEMPERATURE OF THE FOOD. Microwave timing is affected by the temperature of the food before it is cooked. Cold food takes longer to cook than room-temperature food. Although this is true for other cooking methods, the short cooking time with microwaves makes it more noticeable.

TURNING. There are two ways of turning food in microwave cooking. Large foods must be turned over during the cooking time. Rotating the pan of food is the second type of turning. Because all microwave ovens cook faster in some spots than others, turning the pan 180 degrees (a half turn) during microwave cooking promotes even doneness. Some microwave ovens come equipped with a turntable that continually rotates the food, eliminating the need to stop the microwave to rotate food manually.

USING METAL. Since metal stops microwaves, food inside a metal pan will not cook. When two pieces of metal are close to each other in a microwave, a static spark, called *arcing,* may occur. This can be caused by a metal utensil being near the oven wall or two pieces of foil next to each other. A metal trim such as a silver or gold rim on a china bowl may become overheated and melt and/or cause an arc between one area of the rim and another.

WRAPPING FOODS. A cover or lid over a container of food prevents splattering. It also retains steam, which encourages even cooking and somewhat reduces cooking time. The steam helps to distribute the heat evenly, speeds cooking, and prevents dehydration. When heavy-duty foodservice plastic wrap is used as a cover, it should be pierced or should only partially cover the container to prevent bursting when steam is created inside the container. Tight-fitting plastic lids also should be partially open during cooking. To prevent spattering but not retain steam, a wax paper or paper towel cover is commonly used.

Cooking Material and Utensils

Many types of paper, glass, ceramic, and plastic containers can be used in the microwave. Metal tools should not be used because they may cause arcing and prevent cooking. Metal handles and metal trim on glass and plastic items should be avoided. Here are suggestions for using various materials and utensils in the microwave.

Type	Uses	Comments
Paper	Absorbing moisture	Recycled paper may have metal flecks.
	Preventing splatters	Avoid paper towels with nylon webbing. Dye from some paper may fade onto food. Paper may catch on fire when used with dry, high-fat foods such as popcorn with oil.
Glass jars	Syrups, toppings	Remove metal cap and any metal bands. Do not heat baby food in jars. Do not heat food in jars to high temperatures. Jars and bottles with small necks may explode if heated too hot or too fast.
Styrofoam	Short-term heating	Food at high temperatures may melt styrofoam.
Plastic wrap	Covering	Wrap foods loosely or vent for cooking with plastic approved for food use.
Boil-in bags	All cooking	Metal twist tie closures may arc and heat to very high temperatures.
Storage bags	Very short-term heating	Storage bags may melt.
Microwave-safe plastic bags	All cooking	Use only recommended closures and vent before microwaving.
Plastic lids	All cooking	Soft plastic lids may warp or melt, and tightly sealed containers may burst or explode.
Glass	All cooking	Not recommended for quantity food-service because of danger of broken glass getting into food.
Specialty microwave pans	Baking	Use plastic bundt and cake pans, plastic muffin tins, and plastic ring molds.
Browning skillet	Meats	Metal in pan is preheated.
Special nonmetal microwave thermometer		Only special thermometers can be used in the oven while it is on.
Temperature probe		Thermometer attaches to oven to measure temperature while cooking. Oven shuts off when food reaches preset temperature.

Here are materials that are *not* recommended for microwave cooking.

Utensil	Comments
Foil-lined paper	Foil will cause arcing and prevent cooking.
Metal	Metal causes arcing and prevents cooking. Do not use glass, plastic, or ceramic with metal screws, handles, or lids.
China	
Corning centura	Dishes may contain metal
Fitz and Floyd oven-to-table	Dishes may contain metal.
Melamine, Melmac	Dishes contain metal.
Lead crystal	Dishes contain metal.

Foods Not Recommended for Microwaving

Foods leavened by steam cannot be cooked in the microwave because the fast cooking does not permit enough steam to develop to make the food rise. Here are types of foods that are not suited to this type of cooking.

Food	Comments
Cakes (angel food, sponge, chiffon)	Hot dry air is needed to set their structure.
Crispy fried foods	Surface cooking or deep fat frying is needed for browning and crispness. Microwavable precooked fried foods can be preheated in the microwave.
Deep-fried foods	Fat will spatter and overheating may result in a fire.
Eggs in shells and whole peeled, hard-cooked eggs	They may burst when cooked or reheated.
Crusty breads	Only reheating is recommended. Dry surface heat is needed for toasting French bread.
Hash-browned potatoes	Hot surface cooking is needed for a crisp, brown surface. Some special frozen potatoes may be microwaved.
Home canning	Traditional jars will not be heated evenly and will probably explode. Canning is never acceptable in quantity foodservice. With special equipment and jars microwave canning is possible. However, little time is saved and nonacid foods should never be canned.
Pancakes, crepes	Surface cooking is needed. Microwave works well for reheating.
Pasta	No time is saved. Foam may boil over pan.
Popcorn	Only special microwave popcorn bags or pan should be used. A brown paper bag may catch on fire if used for popping corn.
Regular rice	No time is saved. Special microwave rice products can be used.
Steak	Hot dry heat is needed. Microwaved meat has steamed flavor and appearance.
Turkey, capon	Needs frequent turning and rotating. It does not brown.

MICROWAVE USES IN QUANTITY FOOD PRODUCTION

In quantity-foodservice operations, a microwave can be used several ways to improve the quality of food served. The most important use is to reheat foods for client meals that have been delayed. When a meal must be reheated, a microwave can be used to serve food with almost no change in quality.

When a meal must be held for a few clients, it is better to refrigerate and then reheat the foods by microwave to 160–180 F to avoid dehydration. Foods dry out and overcook when they are held hot. Table 11.2 on pages 180 and 181 lists reheating times for prepared foods. The times should be correct in 700- to 825-watt ovens. The instruction book for the microwave oven should also have suggested reheating times.

Another use for the microwave is cooking or heating ingredients for recipes that will be cooked by microwaving or traditional cooking methods. Table 11.3 on pages 182 and 183 gives directions for preparing ingredients, such as toasting nuts, melting butter or chocolate, and precooking small quantities of vegetables.

Many foods can be thawed in small quantities in the microwave oven. Table 11.4 on pages 183 and 184 gives directions for thawing raw and cooked foods. In quantity production, using microwaving for thawing is appropriate only for individual portions of raw foods or for a few portions of cooked, frozen foods. Care must be used in thawing any food to prevent contamination and growth of microorganisms and a reduction in quality. Although meat and poultry can be thawed in the microwave, this procedure should only be used in an emergency and following the directions in Table 11.4 exactly. The preferred method for thawing any raw foods is in a refrigerator.

MICROWAVE SAFETY

Important safety precautions must be followed when using microwaves. A regular schedule of cleaning inside the oven and door gasket has to be maintained. The oven and door become coated with food residue that can become spoiled or rancid. Fat residue may smoke and burn. The door gasket of microwave ovens must be closed before the oven will turn on, and a dirty door gasket will prevent the door from closing completely.

The facility's maintenance department should regularly check the oven for leakage of microwaves. The concentrated microwaves that could escape from a leaking door not only lower the efficiency of the oven but can also cause serious burns.

In a microwave oven, food can be heated to very hot temperatures in a short time. Microwaved foods that are overheated usually do not look different from those that are a safe temperature to serve. Checking the temperature of foods with a thermometer before serving should be a requirement when microwaves are used in nursing facilities. Because microwaved foods have hotter and cooler spots, the temperature should be checked in more than one location in the food. Stopping the cooking several times to stir the food will help make the temperature the same throughout.

Hot-spotting (the unequal heating of foods) makes it unsafe to use a microwave to cook fresh pork. When meat is cooked medium to well-done, some parts may not reach 140 F. Because this temperature is not high enough to kill trichina, pork should not be cooked in the microwave.

SUMMARY

Microwave ovens use short radio waves called microwaves to produce heat in foods by friction of molecules. These microwaves are attracted to the moisture, sugar, and fat in foods. Food begins to cook near the surface, then continues when heat is conducted throughout the food.

Foods that are small, moist, and uniform in size and shape microwave best. As the size and/or quantity of food increases, the cooking time increases.

In foodservice operations, microwaves are most useful for reheating foods, cooking small quantities, and melting fats.

Table 11.2. Microwave reheating times for common foods in a 700- to 825-watt microwave oven

Food	Amount	Power Level	Approximate Time (min)
Appetizers			
Meatballs, riblets, cocktail franks, etc., ½-c serving	1–2 servings	Hi (10)	1½–4
	3–4 servings	Hi (10)	4–6
Dips, cream or process cheese	½ c	Med (5)	2½–3½
	1 c	Med (5)	3–5
Small pizzas, egg rolls, etc.	2–4 servings	Hi (10)	1–2

Tip: Cover saucy appetizers with wax paper. Cover dips with plastic wrap. Do not cover pastry bites or they will not be crisp.

Food	Amount	Power Level	Approximate Time (min)
Preplated foods			
Meat plus 2 vegetables	1 plate	Hi (10)	2–4

Tip: A temperature probe or microwave-safe thermometer works well in saucy dishes or vegetables (use in largest serving) but not in meat slices. Cover plate of food with waxed paper or plastic wrap.

Food	Amount	Power Level	Approximate Time (min)
Meats and entrees			
Chop suey, spaghetti, creamed chicken, chili, stew, macaroni and cheese, etc., ¾- to 1-c serving	1–2 servings	Hi (10)	3–7
	3–4 servings	Hi (10)	8–14
	1 16-oz can	Hi (10)	4½–6
Thinly sliced roasted meat, 3- to 4-oz serving			
Rare	1–2 servings	Med-hi (7)	1–2
	3–4 servings	Med-hi (7)	2–3½
Well-done	1–2 servings	Med-hi (7)	1½–3
	3–4 servings	Med-hi (7)	3–5
Steaks, chops, ribs, other meat pieces			
Rare	1–2 servings	Med-hi (7)	2½–4
	3–4 servings	Med-hi (7)	5–9
Well-done	1–2 servings	Med-hi (7)	2–3
	3–4 servings	Med-hi (7)	4–7
Hamburgers or meat loaf, 4-oz serving	1–2 servings	Hi (10)	¾–2
	3–4 servings	Hi (10)	1½–3½
Chicken pieces	1–2 pieces	Hi (10)	¾–2
	3–4 pieces	Hi (10)	2–3½
Hot dogs and sausages	1–2	Hi (10)	1–2
	3–4	Hi (10)	2½–3½
Rice and pasta, ⅔- to ¾-c serving	1–2 servings	Hi (10)	1–2
Topped or mixed with sauce, ⅔- to ¾-c serving	1–2 servings	Hi (10)	3–6
	3–4 servings	Hi (10)	8–12

Tip: Cover saucy main dishes with plastic wrap. Cover other main dishes and meats with waxed paper. Do not cover rare or medium rare meats.

Food	Amount	Power Level	Approximate Time (min)
Sandwiches and soups			
Moist filling: sloppy joe, barbecue, ham salad, etc., in bun, ⅓-c serving	1–2 servings	Med-hi (7)	1–2½
	3–4 servings	Med-hi (7)	2½–4
Thick meat-cheese filling, with firm bread	1–2 servings	Med-hi (7)	2–3
	3–4 servings	Med-hi (7)	4–5
Soup			
Water-based, 1-c serving	1–2 servings	Hi (10)	2–6
	3–4 serving	Hi (10)	7–11
	1 10-oz can	Hi (10)	6–7
Milk-based, 1-c serving	1–2 servings	Med-hi (7)	3–8
	3–4 servings	Med-hi (7)	10–14
	1 10-oz can reconstituted	Med-hi (7)	7–8

Tip: Use paper towel or napkin to cover sandwiches. Cover soups with waxed paper or plastic wrap.

Table 11.2. (continued)

Food	Amount	Power Level	Approximate Time (min)
Vegetables, cooking			
Small pieces of peas, beans, corn, etc.,	1–2 servings	Hi (10)	1–3
½-c serving	3–4 servings	Hi (10)	3–4½
	1 16-oz can	Hi (10)	3½–4½
Whole or large pieces of green beans,	1–2 servings	Hi (10)	1½–3
asparagus spears, corn on the cob, etc.	3–4 servings	Hi (10)	3–4½
	1 16-oz can	Hi (10)	4–4½
Mashed, ½-c serving	1–2 servings	Hi (10)	1–3
	3–4 servings	Hi (10)	4–7

Tip: Cover vegetables for the most-even heating.

Sauces, warming			
Dessert: chocolate, butterscotch, etc.	½ c	Hi (10)	½–1½
	1 c	Hi (10)	1½–2½
Meat, or main dish, chunky type: thick gravy,	½ c	Hi (10)	1½–2½
spaghetti sauce, etc.	1 c	Hi (10)	2½–4
	1 16-oz can	Hi (10)	4–6
Starch-thickened sauces	½ c	Med (5)	2
White sauce	½ c	Hi (10)	1–1½
	1 c	Hi (10)	2–2½

Tip: Cover food to prevent spatter.

Bakery foods, warming from room temperature			
Cake, coffee cake, doughnuts, sweet rolls,	1 portion	Low (3)	½–1
nut or fruit bread, etc.	2 portions	Low (3)	1½–2
	4 portions	Low (3)	1½–2½
	Whole 9″ cake or 1 doz rolls or doughnuts	Low (3)	2–4
Dinner rolls, muffins, etc.	1	Med (5)	¼–½
	2	Med (5)	½–¾
	4	Med (5)	½–1
	6–8	Med (5)	¾–1½
Pies: fruit, nut, or custard, ⅛ of 9″ pie	1 portion	Hi (10)	½–1
	2 portions	Hi (10)	1–1½
	4 portions	Med-hi (7)	2½–3½
	9″ pie	Med-hi (7)	5–7

Tip: Do not cover. Use ½ min for custard pie.

Grilled foods, reheating			
Pancakes, French toast or waffles, 3×4″ serving			
Plain, no topping	2–3 pieces	Hi (10)	½–1
Syrup and butter	2–3 pieces	Hi (10)	1–1¼
With 2 sausage patties (cooked)	2–3 pieces	Hi (10)	1¼–1½

Tip: Do not cover.

Beverages, heating			
Coffee, tea, cider	1–2 cups	Hi (10)	1½–3½
Other water-based	3–4 cups	Hi (10)	6–7
Cocoa, other milk-based	1–2 cups	Med-hi (7)	2½–7
	3–4 cups	Med-hi (7)	7–10

Tip: Do not cover.

Table 11.3. Suggestions for using a 700-watt microwave to prepare recipe ingredients

Ingredient	Instructions
Toasting nuts	Place ½ c of nuts in shallow bowl. Microcook, uncovered, on 100% power about 3 min or until toasted, stirring frequently. Repeat as needed for recipe.
Blanching almonds	In small nonmetal bowl microcook 1 c of water, uncovered, on 100% power for 2–3 min or until boiling. Add ½ c of almonds to water. Microcook, uncovered, on 100% power for 1½ min. Drain, rinse almonds with cold water, and slip off skins.
Toasting coconut	Place flaked or shredded coconut on a plate. Microcook, uncovered, on 100% power until light brown, stirring every 20 sec. Allow 1–1½ min for ¼ c or 1½–2 min for ½ c.
Warming ice cream toppings	Spoon topping into bowl. Microcook, uncovered, on 100% power until warm, allowing about 15 sec for 2 TBSP, about 25 sec for ¼ c, or about 45 sec for ½ c.
Softening ice cream	Microcook 1 pt solidly frozen ice cream, uncovered, on 100% power about 15 sec or until soft enough to serve.
Rehydrating dried fruit	In 2-c glass measure microcook 1 c of water, uncovered, on 100% power for 2–3 min or until boiling. Stir in ½ c of dried fruit. Let stand for 5–10 min and drain. Repeat as needed for recipe.
Softening butter or margarine	Unwrap butter or margarine and place in small dish. Microcook, uncovered, on 10% power, allowing about 30 sec for 2 TBSP or 50 sec to 1 min for ¼ c.
Melting butter or margarine	Unwrap butter or margarine and place in bowl. Microcook, uncovered, on 100% power, allowing 25–30 sec for 2 TBSP or about 40 sec for ¼ c.
Softening cream cheese	Unwrap one 3-oz package of cream cheese and place in small bowl. Microcook, uncovered, on 30% power about 1 min or until soft.
Melting chocolate squares	Unwrap chocolate and place in small bowl. Microcook, uncovered, on 100% power until melted, stirring once. Allow 1½–1¾ min for one 1-oz square, or 1¾–2 min for two 1-oz squares.
Melting chocolate pieces	In bowl microcook chocolate pieces, uncovered, on 100% power until melted, stirring once. Allow 1–1½ min for ½ of 6-oz package (½ c) or 1½–2 min for one 6-oz package (1 c).
Melting confectioner's coating	In small bowl microcook confectioner's coating, uncovered, on 100% power until melted, stirring once. Allow 1–1¼ min for one 2-oz square or about 1½ min for two 2-oz squares.
Melting caramels	Unwrap caramels and place in 1-c glass measure. Microcook, uncovered, on 100% power until melted, stirring once. Allow 45 sec to 1 min for 14 caramels (about ½ c) or 1–1½ min for 28 caramels (about 1 c).
Flaming liqueur	Place 2 TBSP of liqueur or liquor (at least 80 proof) in 1-c glass measure. Microcook, uncovered, on 100% power for 20 sec. Ignite and pour over food as recipe instructs.
Peeling tomatoes	In 2-c glass measure heat 1 c of water, uncovered, on 100% power for 2–3 min or until boiling. Spear 1 tomato with long-tined fork and submerge it into the hot water; hold about 12 sec. Place tomato under cold running water and slip off peel.
Peeling peaches	In 2-c glass measure heat 1 c of water, uncovered, on 100% power for 2–3 min or until boiling. Spear 1 peach with long-tined fork and submerge it into the hot water; hold about 12 sec. Place peach under cold running water and slip off peel.
Making croutons	Spread 2 c of ½" bread cubes in shallow baking dish. Microcook, uncovered, on 100% power for 3½–4½ min or until crisp and dry, stirring every 2 min.

Table 11.3. (continued)

Ingredient	Instructions
Crisping snacks	Spread 1 c of stale chips, pretzels, crackers, or other snacks in 7″ nonmetal pie plate or shallow baking dish. Microcook, uncovered, on 100% power for 30–45 sec. Let stand 1 min.
Precooking vegetables	Place raw vegetable pieces of uniform size (carrots, zucchini, onions, new potatoes, cauliflower, asparagus, broccoli) in baking dish. Add 1 TBSP of water. Microcook, covered, on 100% power until almost tender. Allow 45 sec to 1 min for ½ c and 1½–2 min for 1 c.
Reheating small meatballs	Place meatballs and sauce in 7″ nonmetal pie plate. Microcook, uncovered, on 100% power until warm, stirring once. Allow 1½–2 min for ½ c or 2½–3 min for 1 c.

Table 11.4. Defrosting frozen foods with a 700-watt microwave oven using 30% power

Food	Step 1 (min)	Step 2 (min)	Instructions
Meat			
Bacon, 1 lb	3½–5	none	Place unopened package in oven. Microwave just until strips can be separated.
4 slices	1–1½	none	
Franks, 1 lb	4–5	none	Place unopened package in oven. Microwave just until franks can be separated.
Beef and pork, ground, 4 oz	1½–3	none	Turn over after step 1.
1 lb	4	4–6	
2 lb	6	6–8	Turn over after step 1. Scrape off softened meat after step 2 and set aside. Break up remaining block and microwave 5–6 min more.
5 lb	12	12–14	Turn over after step 1. Scrape off softened meat after step 2 and set aside. Break up remaining block and microwave 11–12 min more.
Beef, roast	3–4 per lb	3–4 per lb	Place wrapped roast in oven. After step 1 turn roast over. After step 2 let stand for 30 min.
Pork, roast	5–6 per lb	5–6 per lb	Place wrapped roast in oven. Turn over after step 1.
Beef, lamb, pork, veal: steaks, chops, cutlets, spareribs	2–4½ per lb	2–4½ per lb	Place wrapped meat in oven. Turn over after step 1. After step 2, separate pieces with table knife and let stand to complete defrosting.
Sausage			
Bulk, 1 lb, <2″ thick	2½	2½–4½	Turn over after step 1.
1-lb roll	2	2–4	Turn over after step 1.
Link, 1–1½ lb	2	1½–2½	No turning needed.
Patties, 12	1	1–2	No turning needed.
Poultry			
Chicken			
Broiler-fryer, cut up, 2½–3½ lb	7–8	7–8	Place wrapped chicken in oven. Unwrap and turn over after step 1. After step 2, separate pieces and place in cooking dish. Microwave 2–4 min more, if necessary.
Whole, 2–3½ lb	9–11	9–11	Place wrapped chicken in oven. After step 1, unwrap and turn chicken over. Shield warm areas with foil.
Cornish hen, 1½–1¾ lb	7–9	6–7	Remove all metal. Place one wrapped hen in oven breast side up. Turn package over after step 1.
Duckling	2–3 per lb	2–3 per lb	Place wrapped duckling in oven. After step 2, unwrap and turn duckling over. Shield warm areas with foil.

Table 11.4. (continued)

Food	Step 1 (min)	Step 2 (min)	Instructions
Turkey, 11 lb	4½–5½ per lb	4½–5½ per lb	Place wrapped turkey breast side down. After step 1, unwrap and turn turkey breast side up. After step 2, run cool water into cavity until giblets and neck can be removed. Let soak in very cool water for 1–2 hr or refrigerate overnight to complete defrosting. If turkey cavity is thawed, turkey should be cooked immediately.
Butter and margarine, ¼-lb stick	1–1½	none	
Fish and seafood			
Fillets	4 per lb	4–6 per lb	Place wrapped fish in oven (if fish is frozen in water, place in baking dish). Rotate ½ turn after step 1. After step 2, hold under cold water to separate.
Fish steaks, 6 oz	2–3	none	
Whole fish, 8–10 oz	2	2–4	Place fish in baking dish. Turn over after step 1. After step 2, rinse cavity with cold water to complete defrosting.
Shrimp, 8 oz	3–5	none	
Shellfish			
Small pieces, 1 lb	7–8	none	Spread shellfish in single layer in baking dish.
Shellfish, solid blocks, 6-oz package	4–5	none	Place block in baking dish.
Oysters, 10-oz can	6½–8	none	Place block in baking dish. Break up with fork after step 1.
Scallop, 1 lb	7–9	none	Place wrapped scallops in oven.
Shellfish, large, whole	4–6	none	Arrange in baking dish.
Crab legs, 1–2, 8–10 oz	4–6	none	
Lobster tail			
1, 6–9 oz	3–5	none	Arrange in baking dish.
2, 6–9 oz	5–8	none	
Bakery products			
Bread or buns, 1 lb	3–4	none	
Bread dough, 1 lb	15–17	none	Use only 10% power.
Heat-and-serve rolls, ½ doz	2–4	none	
Bread, 2 slices	20–40 sec	none	
Coffee cake, 11–15 oz	3½–5	none	
Coffee ring, 10 oz	3½–4	none	
Sweet rolls, 8–12 oz	2½–4	none	
Doughnuts, 1–3	1½–4½	none	
Doughnuts, glazed, 1 doz	2–3	none	
French toast, 2 slices	5½–6½	none	
Cake, whole, frosted, 2–3 layers, 1 lb	2–2½	none	
Cake, filled or topped, 1 layer, 12–16 oz	2–3	none	
Pound cake, 11 oz	2	none	
Cheesecake, plain or fruit top, 1 lb	5–7	none	
Crunch cakes and cupcakes	½–1 each	none	
Fruit or nut pie, 8″	8–10	none	
Cream or custard pie, 14 oz	2–3	none	
Fruit			
IQF, 10–16 oz	6–11	none	Place unopened package in oven with foil or metal removed. After step 1, break up with fork. Repeat if necessary.
Plastic pouch, 1–2, 10-oz pkg	5–10	none	Place unopened package in oven. Flex package once before cooking.

LEARNING ACTIVITIES

Activity 1: Finding the Wattage of a Microwave Oven

1. Follow these steps.

    A. Put 1 cup 40–50 F water and 8 large (1-inch) ice cubes in a quart measure.

    B. Stir the water and ice cubes for exactly 1 minute.

    C. Pour 1 cup of the cold water into a microwave-safe quart measure. Discard the remaining ice cubes and water.

    D. Place the water in the microwave. Cook uncovered at the highest power level until a full rolling boil is reached.

    E. Record the minutes and seconds of cooking time needed for the water to reach the full boil.

2. Compare the time with the chart in Table 11.1 to find out the wattage of the oven and conversion factors.

3. Calculate the approximate cooking time for your oven when the recipe for a 700-watt oven lists the cooking time as 6 minutes.

    6 minutes × ____ (adjustment factor) = ____ minutes and ____ seconds

Activity 2: Evaluating Microwave-safe Equipment

1. Make a list of your facility's pans, containers, and utensils that could be safely used in microwave cooking.

Activity 3: Comparing Two Potatoes Microwaved in Different Wrappings

1. Bake 2 potatoes in the microwave oven, following these steps.

    A. Choose 2 baking potatoes of the same size and shape. Scrub them and poke three holes in each with a sharp knife.

    B. Wrap 1 potato in plastic wrap. Wrap the other in a white paper towel.

    C. In a 650-watt oven, cooking 1 potato requires 4 minutes, and 2 potatoes 7 minutes. Find the correct cooking time for your microwave. Calculate the adjustment factors using Table 11.1.

    D. Bake the potatoes one at a time in the microwave.

    E. When the oven shuts off remove the potatoes.

2. Without unwrapping the potatoes, insert a thermometer into the center of each and record the temperature on the chart.

Activity 3, Steps 2 and 3

	Temperature Right Out of Oven	Temperature after 3 Min
Plastic-wrapped potato		
Paper-wrapped potato		

3. Allow the potatoes to stand for 3 minutes and then record the temperatures again on the chart. Did the temperatures change? Why?

4. Cut both potatoes in half. What are the differences in texture and color of the skin and interior? What caused these differences?

REVIEW QUESTIONS

Short Answer

1. Define the following words.

    A. Arcing

    B. Wattage

    C. Shielding

    D. Conduction

    E. Density

2. Give at least two reasons why it is important to check the temperature of microwaved foods.

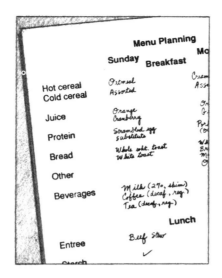

12. MENU PLANNING

A *menu* is a detailed list of foods to be served at a meal or a list of items offered by a facility's foodservice department. Good menu planning will provide nutritious meals that clients enjoy. The menu largely determines client satisfaction, which in turn leads to the success of the foodservice department. Because many facilities have limited resources of time, money, equipment, and personnel, it is necessary to plan menus carefully in order to meet the psychological, social, nutritional, and aesthetic needs of the clients.

The menu controls purchasing and production. It determines the personnel needed, their work schedules, and costs. In effect, it provides the framework for the foodservice department's budget.

FACTORS AFFECTING MENU PLANNING

It is important to consider the following when planning menus:

- The clientele—their nutritional needs, habits, and preferences
- The availability of food
- The availability of equipment and the physical arrangement of the kitchen
- The availability of personnel skills
- The aesthetic appearance of food to the client
- The style of service
- The money budgeted for food

Clientele

In a nursing facility a primary consideration is the health of the clients. The menu must provide three meals and appropriate snacks every day for a nutritionally balanced diet. Modified versions for special needs, such as low-fat, low-sodium, liquid, or calorie-controlled diets also must be provided. In addition, consideration should be given to chewing difficulties and limited mobility and activity.

In addition, it is important for the person planning the menu to be aware of the problems peculiar to clients of nursing facilities. They have fixed food habits and preferences that have developed over many years. They also have different cultural and ethnic backgrounds, some of which result in certain food likes and dislikes. Careful consideration of racial and regional foods and customs and religious restrictions will result in a menu that is well accepted by the clients.

In planning menus it is of primary importance to meet the nutritional needs of the clients, especially in nursing facilities that provide their clients' total nutritional requirements. It is necessary to provide the kinds and amounts of food that will help keep clients in a healthy state. Because older individuals generally need fewer calories than when they were young, the calories they do eat must be nutrient-dense, not "empty" calories.

Protein, carbohydrates, fats, minerals, vitamins, and water are necessary nutrients. They are needed to build and repair body tissues, furnish energy for heat and activity, and regulate body functions. Here are the key nutrients needed by the body, with their functions and sources.

Key Nutrient	Functions	Sources
Protein	Builds and maintains all tissues Forms an important of enzymes, hormones, and body fluids Helps form antibodies to fight infection Supplies energy	Top-quality proteins for tissue building and repair: lean meat, poultry, fish, seafood, eggs, milk, and cheese; next best proteins: dry beans, dry peas, nuts; lower-quality proteins: cereals, bread, vegetables, fruits
Calcium	Builds bones and teeth Helps blood to clot Help nerves, muscles, and heart to function properly	Milk (whole, low-fat, skim, buttermilk), fresh, dried, canned; cheese, especially cheddar types; ice cream, ice milk; leafy vegetables (collard, dandelion, kale, mustard, turnip greens)
Iron	Combines with protein to make hemoglobin, the red substance of blood that carries oxygen from the lungs to muscles, brain, and other parts of the body Helps cells use oxygen	Liver, kidney, heart, oysters, lean meat; egg yolk; dry beans, dry peas; dark-green leafy vegetables; dried fruits; whole-grain and enriched breads and cereals; molasses
Iodine	Helps thyroid gland to work properly	Iodized salt; saltwater fish, other seafood
Vitamin A	Helps eyes adjust to dim light Helps keep skin smooth Helps keep lining of mouth, nose, throat, and digestive tract healthy and resistant to infection Promotes growth	Liver; dark-green and deep-yellow vegetables (broccoli, turnip and other leafy greens, carrots, pumpkin, sweet potatoes, winter squash); apricots, cantaloupe; butter, fortified margarine
Thiamin	Helps body cells obtain energy from food	Lean pork, heart, kidney, liver; dry beans, dry peas; whole-grain and enriched cereals and breads; some nuts

Key Nutrient	Functions	Sources
Ascorbic acid (Vitamin C)	Helps hold body cells together and strengthens walls of blood vessels Helps in healing wounds Helps tooth and bone formation	Cantaloupe, grapefruit, oranges, strawberries; broccoli, Brussels sprouts, raw cabbage, collards, green and sweet red peppers, tomatoes; mustard and turnip greens; potatoes cooked in jackets
Riboflavin	Helps body cells use oxygen to release energy from food	Milk, liver, kidney, heart, lean meat; eggs; dark leafy greens
Niacin	Helps body cells use oxygen to produce energy Helps to maintain health of skin, tongue, digestive tract, and nervous system	Liver, lean meat, poultry, fish; peanuts and peanut butter; dry beans, dry peas; whole grain and enriched breads and cereals
Vitamin D	Helps body use calcium and phosphorus to build strong bones and teeth, especially important in growing children and during pregnancy and lactation	Fish liver oils; foods fortified with vitamin D (milk); direct sunlight (produces vitamin D from cholesterol in skin)
Carbohydrates	Supply food energy Help body use fat efficiently Spare protein for purposes of body building and repair	Starches: breads, cereals, corn, grits, potatoes, rice, spaghetti, macaroni, noodles; sugars: honey, molasses, syrups, sugar, other sweets
Fats	Supply food energy in compact form (weight for weight supplies twice as much energy as carbohydrates and proteins) Supply essential fatty acids Help body use certain other nutrients	Cooking fats and oils, butter, margarine, salad dressings, and other oils
Water	Important part of all body cells and fluids Carrier of nutrients to and waste from body cells Aids in digestion and absorption of food Helps to regulate body temperature	Water, beverages, soup, fruits, vegetables (most foods contain some water)

Other nutrients are needed by the body for good nutrition in addition to the key nutrients. However, research shows that if the individual receives enough of the key nutrients, the other nutrients also will be adequate.

The nutrient requirements can be met in many ways. Table 12.1 shows daily food choices developed by the U.S. Department of Agriculture (USDA). It can be used to make sure clients are getting the proper nutrients.

Although certain foods are grouped together, all foods in one group are not exactly the same in nutrients. For this reason, it is necessary to plan a variety of foods in the menu to make sure the client receives all the nutrients needed over time. Almost everyone should have at least the minimum number of servings from each food group.

Table 12.1. Daily food choice pattern

Food Groups	Suggested Daily Servings
Breads, cereals, other grain products, whole-grain or enriched	6–11 (including several servings a day of whole-grain products) 1 serving: 1 slice bread or ½ English muffin, ½ hamburger bun, ½ c ready-to-eat cereal, rice, pasta
Fruits: citrus, melon, berries, others	2–4 1 serving: 1 apple, banana, orange; ½ grapefruit; ¾ c juice; ½ c berries, cooked or canned fruit; or ¼ c dried fruit
Vegetables: dark-green leafy, deep-yellow, dry beans and peas (legumes), starchy, others	3–5 (including all types regularly and dark-green leafy vegetables and dry beans and peas several times a week) 1 serving: ½ c cooked or chopped raw vegetable or 1 c leafy raw vegetables
Meat, poultry, fish, alternates (eggs, dry beans and peas, nuts, seeds)	2–3 servings (total 5–7 oz) 1 serving: 2–3 oz boneless cooked lean meat or equivalent (1 oz lean meat = 1 egg; ½ c cooked dry beans, peas, lentils soybeans; or 2 TBSP peanut butter)
Milk, cheese, yogurt	2 servings 1 serving: 1 c milk, 8 oz yogurt, or 1½ oz natural cheese
Fats, sweets, alcoholic beverages	Avoid too many fats and sweets. Drink any alcoholic beverages in moderation.

Source: Adapted from *Dietary Guidelines for Americans,* Iowa State University Extension, Ames. Reprinted 1989.

The seven Dietary Guidelines for Americans developed by the USDA should be considered when planning menus. Menus should include only a moderate amount of fat and cholesterol, sugar, and sodium. Foods with adequate starch and fiber are important. Here are guidelines that will help menu planners meet these concerns:

Guideline	Reasons	Suggestions
Use fat and cholesterol in moderation.	It raises the blood cholesterol level.	Choose lean meat, fish, poultry, dry beans and peas; limit intake of butter, cream, hydrogenated margarine, shortenings, and coconut oil; broil, bake, or boil food rather than fry.
Avoid too much sugar.	It provides only calories, lacks nutrients, and promotes tooth decay.	Use less of all sugars, including honey and syrups; use fewer sweet foods such as ice cream, cookies, cakes, and candy; select fresh fruits or fruits canned in their own juice or in light syrup; use fresh fruits for dessert.

Guideline	Reasons	Suggestions
Avoid too much sodium.	It increases the risk for hypertension.	Cook with only small amounts of or no added salt; use spices or herbs for flavoring; limit salty foods, salted nuts, potato chips, condiments, sauce, garlic salt, process cheese, pickled foods, and cured or processed meats.
Include adequate fiber and starch.	They promote normal bowel movements.	Use fruit for dessert in place of pie and cake; select sources such as whole-grain breads and cereals, vegetables, fruits, dry beans, and dry peas.

Availability of Food

A limiting effect on the menu may be the nearness of the source of foods, especially fresh foods, to the facility. Menu planners must use their knowledge of seasonal foods to include them on the menu when they are at their peak of quality and lowest cost. It is important to be aware of new foods on the market that might provide interest by adding variety and quality to the menu items offered.

Availability of Equipment and Arrangement of Physical Facilities

Menus must be planned with the kitchen and serving area arrangements in mind to be successful. The menu plan for the day must be planned to be produced with the available workspace and equipment. For example, planning many foods for one meal that must be cooked in the oven will cause an overload on the ovens and may result in production conflicts. Likewise, it is hard to use many items in one day that need refrigerator or freezer space. In addition, the amount of freezer and refrigerator space must be considered when determining delivery of food to the facility.

Personnel Skills

When menu planners understand the relationship between the variety and complexity of menu items and the number of personnel needed, fewer problems result. Some menu items require preparation, while others must be prepared just before serving. Caution needs to be taken to prevent too many last-minute-preparation tasks. Personnel skills will also influence the foods that can be prepared. If no one has the skill to make pastries, the planner should not include pies made from scratch on the menu.

Money Budgeted for Food

In nursing facilities, a raw food cost allowance per person per meal per day may be used to determine the amount allowed in the budget for food, labor, and operating expenses. A menu planner must be aware of this allowance as menus are written. It is not necessary to have all meals fall below or at the budgeted amount each day, but over the week or month the average spent for food needs to be in line with the allowance given. This can be done by balancing higher-cost meals with lower-cost meals, which also adds variety to the foods offered to the clients.

When a choice of entrees is offered, it is a good idea to offer two less expensive items at the same time. Likewise, another menu choice should be between two more expensive entrees. When a more expensive entree is offered along with a less expensive one, there is no control over balancing costs, as most clients may select the more expensive entree.

Style of Service

The style of service may determine items that can and cannot be selected for the menu and the form in which some items appear.

THE CYCLE MENU

A *cycle menu* is a set of carefully planned menus rotated according to a definite pattern. After a preplanned period of time, the menu repeats itself. In a nursing facility, a longer cycle is normally used because of the length of stay of clients. The menu is different each day during the cycle, with many of the same foods used but in different combinations and on different days. There may be a different cycle for each season of the year or modifications may be made to the main cycle for winter and summer.

Before planning menus, a decision needs to be made about the length of the cycle menu. Five weeks is a popular length for a cycle menu in a nursing facility.

A nursing facility generally uses a *nonselective* cycle menu, which means the client does not select foods. Client acceptance may be higher, however, if some modification is made to the nonselective menu. Choices may be provided through a soup and salad bar; by offering some foods each day, such as fresh fruit, cottage cheese, and soup; and by offering a choice between two entrees at one meal or when a less popular entree is offered. It is important to have a record of individual preferences on file to better meet each client's needs.

Advantages of the Cycle Menu

A cycle menu has several advantages for the menu planner, especially if labor and equipment are considered during the planning time.

- Time is available to revise menus to meet changes in holidays, personnel, and season of the year; to plan the use of leftovers; to plan the use of new food and recipes; and to take advantage of lower-cost foods.
- Time is available for supervision of personnel.
- Recipes and preparation procedures are standardized.
- Equipment use is more efficient.
- Workloads are more evenly distributed and improved with repeated use.
- Forecasting and purchasing are simpler and better.
- Money is saved because unpopular foods are eliminated and the amount of food actually needed is planned and purchased.

Disadvantage of the Cycle Menu

The main disadvantage of cycle menus is that the menus can become monotonous if the cycle is too short or the same food is served on the same day each week. This disadvantage can be easily eliminated by lengthening the cycle and checking to ensure that the same menu is not served on the same day of the week.

PROCEDURES FOR GOOD MENUS

Planning Menus

Good menu planning gives the best results if approached in a systematic way. These are a few ideas that have worked well for many menu planners.

- Plan or revise a menu far enough in advance to have reliable food delivery and to use time efficiently.
- Have a regularly scheduled, undisturbed time for menu planning and revising of the cycle menu so menu planning is a part of a regular routine.
- Plan at a specific place, so necessary materials can be left there.
- Have standardized recipes, menu and idea files, and suggestions from clients available to provide ideas.
- Have the needed supplies nearby, such as the telephone and menu forms.
- Review the menus used during the past 2 weeks to avoid repetition and monotony.
- Know the market and what foods are in season and plentiful, because they usually have better quality and probably are the better buy.
- Be alert to new products, trends in client preferences, and menu items that are well accepted.
- Review the records of food on hand and take inventory, if necessary.
- Add new foods to menus at least once a week to provide variety for the clients and for the foodservice personnel, as well. When a cycle menu is used, add new items when revising it.
- Provide a recipe for all food items, if needed, as well as a time schedule for cooking vegetables and meats.

WHAT TO CONSIDER IN PLANNING MENUS. When planning menus, it is important to visualize how the food will look on the plate or tray, the *presentation* of food. A sense of how flavors will combine and of the contrasts in texture, shape, and consistency is important. Some general considerations can help make a meal appealing.

Color. Color adds interest. One or two bright-colored foods should be included in each meal, not all dark or all light foods. Foods of the same color at one meal, such as a pork chop, creamed potatoes, cauliflower, a pear salad, and white cake are not appealing. Fruits and vegetables, properly prepared, have bright colors that serve as accents. Foods that clash or have unappealing color schemes, such as beets, strawberries, and tomatoes, should not be used in the same meal.

Texture. *Texture* refers to the structure of a food, how it feels to the mouth. The texture of foods can be crisp, soft, chewy, smooth, hard, or grainy. Fresh vegetables provide a contrasting texture to baked chicken and mashed potatoes. On the other hand, cream sauces on entrees do not offer a contrast to cream soups. Here are examples of foods that offer variety in texture.

Food	Texture
Salads, relishes, toast, crusty rolls, crackers	Crispness
Meats	Chewiness
Breads, fish, casseroles	Softness
Puddings, sauces	Smoothness

Temperature. The temperatures at which foods are commonly served are hot, cold, and room temperature. A meal with foods of various temperatures is more interesting.

Consistency. *Consistency* is the way foods stick together, the degree of firmness, density, or flow of a product. It is referred to as firm, thin, or thick. The consistency of a meal of Swiss steak with gravy, scalloped potatoes, and creamed carrots and peas could be varied by substituting mashed potatoes and buttered carrots and peas.

Size, Shape, and Form. Interest can be created in a menu by varying the forms in which the food is served. Cutting vegetables in different shapes can add variety and eye appeal. It is important to include whole pieces of food as well as cut vegetables.

Flavor Combinations. Plan a combination of flavors: sweet, sour, salty, and bitter. Strong and mild foods or spicy, highly seasoned and bland foods should be offered together. Flavors should not be duplicated within a meal, such as all tart foods or all sweet foods. Candied sweet potatoes, pecan pie, and orange glazed rolls provide a duplication of sweet foods. If tomato soup is served, tomatoes should not be used as a garnish or in a salad.

Light and Heavy Foods. The balance of light and heavy foods served together, as well as within one meal, should be considered. For example, a heavy cream soup balances out a lighter broiled fish.

Preparation Methods. Several foods prepared the same way will not add variety to the meal, for example, several foods that are creamed, fried, or served with a sauce. Stir-frying foods is a good way to add variety to the traditional baking, broiling, steaming, boiling, frying, and braising.

Presentation. The menu planner should visualize how the food will look on the plate. Arrangement of the food on the plate is just as important as how the food is placed in front of the client. A haphazard arrangement on a plate can turn off the appetite. Consider the following:

- Will the food fit on the plate?
- Is the food placed on the plate for the client's convenience?
- Are portion sizes of different food items in proportion to one another?
- Is there height on the plate rather than a flat presentation?
- Does the food look like it belongs together?

After the menus have been written, the menu planner should decide if a garnish is necessary. If there is enough color on the plate, then there may be no need for extra color. Unnecessary garnishes are costly. If garnishes are used, they should be simple and appropriate to the foods on the plate and should be edible.

Writing Menus

The following steps are one approach to writing menus in an orderly manner.

1. Write the meat or main dishes for lunch and supper throughout the cycle. Because the entree tends to be the most expensive item, menus are usually planned around it. Popular items may be repeated more than once during the cycle.
2. Select the vegetables. If a sauce is used with the entree, a plain vegetable or combination of vegetables may be chosen.
3. Select the salad and breads.
4. Select the dessert.
5. Write the breakfast menu. Writing breakfast last helps the planner to fit all the necessary

nutrients into the day's meals. A decision to include protein in the breakfast can be based on the protein in step 1.
6. Include at least two foods in each meal that can be used for soft diets and with little change for other modified diets.
7. Check menus for repetition to be sure that one food is not used more than once on the same day, 2 days in a row, or on the same day each week.
8. Evaluate the menus, using a checklist.

The following checklist will help the menu planner determine if all the factors needed for good planning have been included.

- Do the menus provide for the nutritional needs of the clients?
- Are the foods in season, available, and within the budget?
- Can the foods be prepared with the equipment and personnel available?
- Do the foods offer contrasts
    - in color?
    - in texture?
    - in temperature?
    - in consistency?
    - in size, shape, and form?
    - in flavor?
    - in lightness and heaviness?
    - in preparation methods?
- Are personnel and equipment workloads balanced?
- Is there a repetition of a food item or flavor within a day or meal?
- Do flavors complement each other?
- Are suitable garnishes and accompaniments used for interest?
- Are new ideas in combinations or preparation methods included?

Evaluating Menus

Menus should be reviewed and changed to improve their acceptability, solve preparation problems, and take advantage of the availability of foods. When the planner records, either on the menu or in a notebook, what was actually served as soon as the food is served, this information can be used to make desirable changes. Such records may include

- Any changes and why they were made
- Balanced use of equipment available for food preparation
- The appearance of tray or plate
- Any last-minute preparation delays

Evaluating menus should be an ongoing process. Each menu should be reviewed before it is served again. Finding out reactions to a menu item helps the menu planner prepare future menus. A foodservice can get this information by

- Observing plate waste
- Surveying the clients for food likes and dislikes
- Using comment cards

SUMMARY

A menu is a detailed list of foods to be served at a meal or a list of items offered by a facility's foodservice department. Menus should be planned to meet clients' psychological, social, nutritional, and aesthetic needs.

Several factors affect menu planning, including the clients, facility resources, and availability of foods. A systematic menu-planning procedure needs to be in place. A cycle menu is a set of carefully planned menus rotated according to a definite pattern. Menus should be evaluated using several criteria. After menus are used, they need to be evaluated as an ongoing process.

LEARNING ACTIVITIES

Activity 1: Planning a Week's Menu

1. Use a general menu plan form to plan meals for 1 week at your facility. Carefully choose acceptable foods and consider factors such as clients' needs, availability of food, and skills of personnel, so that the week's menu can be used at a later date. Follow the menu writing procedures presented in this chapter.

2. Evaluate your week's menu.

    A. Evaluate your menu for variety by filling in the chart.

    Activity 1, Step 2A

Is There Variety in the Following?	YES	NO
Color		
Texture		
Temperature		
Consistency		
Size, shape, and form		
Flavor combinations		
Balance of light and heavy foods		
Preparation methods		
Repetition of food		

    B. Evaluate your week's menu for nutritional adequacy. Have any basic nutrients been left out? Which ones?

    _____

    _____

MENU PLANNING 197

Activity 2: Evaluating Menus in Your Facility

1. Look at the same day of the week for each week of the menu cycle. Then look at two weeks' menus in a row.

2. What changes do you recommend?

   _____
   _____
   _____
   _____

Activity 3: Evaluating a Menu

1. Evaluate the following menu for variety using the criteria in the chart in Activity 1.

   Cream of broccoli soup
   Pot roast with gravy
   Mashed potatoes
   Harvard beets
   Cloverleaf rolls
   Vanilla ice cream
   Coffee

2. What changes would you make to meet the criteria?

   _____
   _____

REVIEW QUESTIONS

True or False

1. The menu controls the foods that are purchased and the production schedule in a facility.

   A. True
   B. False

2. When planning menus, it is equally important to consider the client's needs for *key* nutrients as well as *non-key* nutrients.

   A. True
   B. False

3. As long as there are foods selected from the basic food groups, the nutritional needs of a client will be met.

   A. True
   B. False

Multiple Choice

4. A menu should be planned to include *minimal* to *moderate* amounts of all nutrients except for

   A. Fat and cholesterol
   B. Fiber
   C. Sugar
   D. Sodium

5. A set of carefully planned menus rotated according to a set pattern is called

   A. A selective menu
   B. A modified menu
   C. A seasonal cycle menu
   D. A cycle menu

6. One approach to writing menus is

   A. To first write the main dishes for lunch and supper throughout the cycle
   B. To first write entire menus for a particular day of the week for the whole cycle
   C. To first write the breakfast menus throughout the cycle
   D. To use no organized approach as long as it is done the same way each time

7. Menu evaluation looks at these factors:

   A. Acceptability of menu by the client
   B. Availability of food
   C. Production problems
   D. Appearance on tray or plate
   E. All of the above

13. FOOD PURCHASING, RECEIVING, AND STORING

PURCHASING

Purchasing is a necessary function in all foodservice organizations. It involves more than just ordering food. Food purchasers must ensure that the right product, in the correct amount, is received at the time needed, for the best price.

Knowledge of food characteristics that indicate quality, how the food was processed, and how that processing affected the quality is essential regardless of who makes purchasing decisions.

Purchasing functions include

- Determining what foods and supplies need to be purchased from purveyors
- Establishing the quantity needed
- Studying the market trends, including supply of food, new forms of food, and price changes
- Negotiating with purveyors and ordering food and supplies
- Receiving products
- Storing products
- Issuing products
- Evaluating purchasing methods

Government Regulations about Food Quality

Government agencies have established standards to ensure the quality, safety, and sanitation of food that is purchased. The federal government provides for control of standards. All food shipped in *interstate* commerce (from one state to another) must meet the requirements of one or more federal laws.

GENERAL STANDARDS. The Food and Drug Administration (FDA) is responsible for enforcing the Federal Food, Drug and Cosmetic Act. This act covers the production, manufacture, and distribution of all foods involved in interstate commerce except meat and poultry. The act forces sellers to offer clean, safe, and sanitary products. It defines *adulteration* (when a food contains substances that are injurious to health, has been held under unsanitary conditions, or contains filthy or decomposed food or portions of diseased animals). The act also defines *misbranding* (when the label does not include the required information or gives misleading information).

Food, Drug and Cosmetic Act standards for food falls in three categories.

- Standards of *identity* define what a food is. They state what must be contained in the product to be labeled by a certain name, such as milk, cheese, margarine, or mayonnaise. The required ingredients, minimum and maximum amounts of each, and optional ingredients are established in these standards.
- Standards of *quality* describe the degree of perfection in shape and uniformity of size. They limit the number and kinds of defects permitted in a product.
- Standards of *fill* regulate the amount of food in a container, which varies among products. False bottoms and slack fill are not allowed.

The Fair Packaging and Labeling Act requires food labels listing common ingredient names. The ingredients must be listed in descending order by weight. Other information, such as the name and address of the manufacturer, packer, or distributor; the quantity; and the names of any additives is also required.

STANDARDS FOR SPECIFIC FOODS. The U.S. Department of Agriculture (USDA) has established uniform standards for state and federally inspected meats, poultry, and eggs that are shipped interstate. They require these products to be inspected for wholesomeness.

Some products that are not shipped across state lines may have to be inspected by state programs with their own standards, some as high as those of the federal inspection program.

Voluntary inspection of fish products and grade standards for fish are under the supervision of the National Marine and Fisheries Service of the Department of Commerce. Only when a product carries a grade designation is there assurance that it has been inspected during processing by federal inspectors.

USDA QUALITY GRADES. The quality standards for most agricultural products have been established by the USDA. These standards refer to the wholesomeness, cleanliness, appearance, size, and texture of the product.

Grades are the market classification of this quality. They show how the product measures against the standard. A product that has been graded under USDA supervision will bear a USDA shield. Grades have been defined for each type of food and vary according to food types. The grades are discussed in the chapters on individual foods.

Brands

A brand is assigned by the company that manufactures a product. The manufacturer's intent is to develop a standard product with its own brand name that will be recognized and demanded by the purchaser. Companies may use the same grading standards as USDA, or they may establish their own standards for their brands, which may result in differences in quality. Regardless of manufacturers' standards, the products bearing the USDA shield still must meet the USDA standards.

The Purchaser's Responsibilities

Food and supplies for foodservice organizations may be purchased by foodservice departments, by purchasing departments, or through cooperative arrangements with other organizations. Whatever method is used, the individual in charge of purchasing for a foodservice department must be involved in all purchasing decisions.

So wise decisions can be made, the purchaser must know

- The federal, state, and local regulations affecting purchasing and how food is marketed
- The quantity and quality standards of the food needed
- The food market, new products, and seasonal fluctuations
- The food preparation methods needed to produce quality products
- How the final product is influenced by the quality purchased
- What storage will be needed for the purchased food
- How to select the most efficient *purveyor* (seller or vendor) for needed products

In making the best decisions for the organization, the purchaser must cooperate with management, employees, and purveyors. The purchaser must be able to communicate effectively with purveyors. The purchaser also must show honesty, maturity, and high ethical standards when acting as the agent for the foodservice operation. To avoid any bias in purchasing decisions, no gifts or favors should be accepted from purveyors. An objective comparison of prices, quality, and service should serve as the only basis for purchasing decisions. The quoted prices are confidential information and should not be shared with competitors, to allow for fair bidding by all interested purveyors.

Centralized Buying

While many foodservice departments purchase their own food and supplies, there are two other purchasing methods, centralized buying and cooperative buying.

Centralized buying occurs when a purchasing department is responsible for getting equipment and supplies for all departments of the organization. This method has been found to be a time-saver for the foodservice department because the foodservice employee does not get involved in the negotiation. However, the foodservice department is usually given the responsibility of buying the fresh produce and other perishables. It is important for the foodservice and purchasing departments to cooperate with each other.

Cooperative Buying

Efforts to reduce costs by buying in volume have led to organizations joining together in cooperative purchasing arrangements. Nursing facilities may join with one or more other nursing facilities, or even other types of foodservice departments, to get better buys. Members of the group are independent organizations with different managements. The group selects an independent buyer and each organization pays a fee for the service. Common specifications and bid schedules must be agreed upon by the cooperating facilities.

Information Needed by Purchasers

Many factors within the foodservice department must be considered when making purchasing decisions.

- The menu: Decisions must be made based on the planned menus rather than planning the menu around purchasing decisions and should include nutritional needs and food habits of clients.
- The number of meals served per day: Past figures can help forecast the total number for all meals and snacks served. These figures should include clients, employees, guests, and visitors.
- The skills and number of personnel: The skills of the personnel determine what can be produced and in what form products must be purchased, for example, whether a cake should be made from scratch or purchased ready-to-serve.
- The available storage: The amount and type of storage will determine the form in which foods are purchased and the amount that can be purchased at one time. If freezer space is unavailable, canned products should be purchased. Consideration also needs to be given to turnover of products, capital investment, and product quality.
- The availability of purveyors: Factors such as location, frequency of delivery, and dependability of purveyors are important.
- Market factors: The season of the year and the market conditions should be considered. Adverse growing conditions can affect food prices. Perishable foods usually have greater price changes during the year than do nonperishable products.
- The food budget: The amount of money available for food will affect the form in which it is purchased. Convenience foods tend to be more expensive than other forms.

Choosing Purveyors

A purveyor is a supplier, usually a wholesaler, who sells primarily to facilities. The success of a purchasing program depends on the suppliers with whom the organization deals. Therefore, the selection of reliable purveyors of products and services is a very important decision. The location of the facility plays a part in the availability of purveyors because there are many more suppliers to choose from in a large metropolitan area.

Factors that should be considered in the selection of purveyors are

- The price
- The product quality
- The product needed: Some items can only be supplied by certain purveyors. The perishability of an item is another limitation that may necessitate using a local purveyor.
- The quality of services, including credit terms, policy on returns, minimum order requirements, variety of goods, delivery schedule, temperature and sanitation conditions of products before delivery, and ability to meet requirements with few substitutions.

The relationship between purchaser and purveyor must be one of trust and confidence. A cooperative attitude, honesty, fairness, and knowledge of foods are important qualities in a purveyor. Past experience, good or bad, can serve as one basis for selection. Staff from other facilities can provide useful information about purveyors.

Methods of Purchasing

Two principal methods of purchasing are informal, either open-market or by informal purchase agreement, and formal, competitive-bid buying. There are other, less common methods, as well. The method used is determined by the organization's policies and procedures.

OPEN-MARKET BUYING. This informal method of buying is more common. Products are ordered from a purveyor based on daily, weekly, or monthly quotations. Daily quotations may be required for fresh fruits and vegetables while a monthly quotation may be used for other grocery items. Price, quality, delivery, and other services are used to determine who receives the order. This method is appropriate when immediate delivery is necessary or when the facility is too small to use formal competitive buying.

INFORMAL PURCHASE AGREEMENT. The informal purchase agreement is used when many products are purchased from a single purveyor. When quantity and needs are determined, the order is placed with the single purveyor. The foodservice receives the bills at a later date. This method often is used when time is an important factor, for example,

- When the time required for a formal purchase order for a small amount is not justified
- When an item can be purchased from one or two purveyors only
- When the need is urgent and immediate delivery is required
- When a formal procedure is not justified because the operation is too small

FORMAL, COMPETITIVE-BID BUYING. With this method, written specifications and quantities needed are submitted to purveyors who are invited to provide quotations. The general conditions include date and method of delivery, methods of payment, willingness to accept all or part of the bid, inspection and certification required for quality, packaging, billing instructions, and date for closing the bids. If a bid is correctly written, there is little chance for error. The submitted bids are opened on a specified date and usually the bidder with the lowest total bid receives the contract.

A particular purveyor may be lower on some items and higher on others, so the decision must be made whether to go with the lower overall bid or to award the bid to several purveyors. Some purveyors may submit quotations for all of the products, and some may submit quotations for a few products. When deciding how to award bids, purchasers should decide if they want to use purveyors who bid only on a few items.

This type of buying lends itself to nonperishable items. However, it requires time and must be done early enough to allow for the process of making the requests and analyzing them. Because of the money involved, there can be a chance of unethical behavior in the purchaser-purveyor relationship.

The contract should be awarded to the most responsible bidder with the best price for the organization. Price alone should not be an indicator. Awarding the contract should be on the basis of price, quality, and service. Consideration must be given to the number of items that are awarded to a specific purveyor.

OTHER TYPES OF PURCHASING. Other methods include cost-plus and prime-vendor or one-stop purchase agreements. With the cost-plus type, the purchaser agrees to buy certain items from the purveyor based on a fixed markup over the purveyor's cost.

Prime-vendor, or one-stop, purchasing is an agreement to buy a majority of certain product categories from one vendor for an agreed-upon price. This method saves time for the purchaser, and the purchaser's overhead, delivery costs, and payment costs are lower than when dealing with several purveyors. There is only one delivery to receive and one bill to pay. However, using a one-stop purveyor has its disadvantages: there is no price competition and if for some reason the purveyor cannot deliver, there will be no supplies.

Food Quality

Decisions must be made about the quality of food best suited to the foodservice operation. Each grade has its use, and no one grade is best for all purposes. For example, in a can of peaches with the highest quality, all the fruit will be the same size, have a uniform deep-yellow color, and contain no visible defects. The peaches will be large with a smooth texture. A lower-quality product may not taste different, but the fruit could be different sizes, vary in color, and have a stringy texture. There are uses for both qualities of peaches in a foodservice operation. The higher quality would be used when peaches are served by themselves. The lower quality is suitable for baking in a pie.

Determining Quantity

The quantity of foods and supplies to be ordered varies with the size of the foodservice operations. The amount to be purchased is determined by

- The number of people being served
- The size of the portion
- The amount of loss through waste and shrinkage during preparation

These needs must be communicated from the production area and the storeroom.

Establishing a minimum and maximum stock order provides a basis for determining how much to purchase. The minimum level is the point below which inventory should not fall. It should include a safety factor (to allow for an unexpected increase in use) and sufficient stock to cover the time between ordering and receiving products. The maximum level is the largest quantity of stock that should ever be in inventory. It is equal to the safety-factor stock plus the estimated usage as determined by past experience and forecasts.

The amount to buy at one time and how often ordering should be done depends on the method of buying, frequency of deliveries, amount of storage space, and nature of the product. Nonperishables are often ordered in large quantities if there is enough storage space because they keep longer. The quantity purchased should minimize quality deterioration and product damage. Perishables are ordered in small quantities because they spoil quickly and must be used within a very short period of time.

In addition, the purchaser should look at quantities that are feasibly economical to purchase. If four cases of a food item are needed for use during 1 month, but the price break is at eight cases, the purchaser may consider buying that quantity provided there is storage space.

Other methods can be used to determine optimal quantities to purchase. The economic order quantity (EOQ) considers order cost, use, item cost, and storage *costs*. This method is appropriate for nonperishable items used in large quantity. The economic order interval (EOI) looks at order cost, use, item cost, and storage *capacity* in determining quantity to purchase.

Writing Specifications

Written specifications for a product are necessary to be sure the desired product is received from the purveyor. They clearly state the exact product, quality, and grade. The specifications should be stated in clear, simple, and concise terms so they are understood by both the purchaser and purveyor.

When specifications for meat are needed, there is a guide that can be used. The Institutional Meat Purchase Specifications (IMPS) system has been developed to simplify and clarify meat purchasing. The system uses numbers to indicate exactly what kind of meat is being ordered.

Specifications are said to be the anchor of a successful purchasing operation. They describe in concrete terms the quality and expected performance of the product to be purchased. When any substitutions are made the specifications allow the foodservice manager to compare them with the product ordered. Each specification should include the following:

- The name of the item: Usually the common or trade name is sufficient, such as frozen green peas.
- The quantity
- The quality grade or brand: Grading has been established to ensure quality standards. Brands also can ensure quality. When there are no grades or brands, the quality description must be very detailed.
- The unit of pricing: This can be a gallon, pound, case, or other unit.
- The packaging, or size of container: Purveyors need to know if the size is based on a 30-pound lug, or a case of No. 6 or No. 10 cans. If a case of 24 heads of lettuce is specified, the weight for the whole case should be stated.
- The count per container or the approximate number per pound: The number of some fruits and vegetables indicates the size of the fruit. For example, a case of 88 oranges means 88 oranges are in 3/5 bushel.

Specific food product information includes the following:

- Fresh fruits and vegetables: variety, weight, degree of ripeness, yield, production area, variety of citrus fruit
- Canned foods: type or style, cut, packing medium, size, syrup density, drained weight, specific gravity (tomato products), percent mixture (fruits in fruit cocktail)
- Frozen foods: variety, sugar or salt ratio, temperature during transportation and delivery, temperature upon receipt of item
- Meats and meat products: age, market class, cut of meat, cutting instructions, trim, weight range, fat content, feed (corn-fed or grass-fed), style of meat product
- Dairy products: milk fat content, milk solids, bacteria count, aging or ripeness (cheese), temperature for delivery and upon receipt, soundness of packaging (no leaking)

Here are sample specifications for baking potatoes:

Potatoes, Irish, fresh
ROP 5 cartons, maximum stack, 15 cartons
No. 1, bakers
50-pound cartons, 100 count
Price by the carton
Shall be Idaho, Washington, or Oregon mature Burbank or more specific gravity. Shall be washed. Count size not to vary more than +/− 1 oz. Shall not have been refrigerated 14 days before delivery.

Purchase Orders

A purchase order is a form the purchaser fills out to place an order for delivery of certain products and supplies. A separate purchase order is written for each purveyor. The purchase order specifies the quantity of each item needed for a specific time period, the unit of measurement, the size and kind of container, the quality desired, and the required delivery date. It includes the names of the foodservice organization, the person making the order, and the signature of the authorized person. The procedure will differ among foodservice operations, but it is always necessary to identify who is authorized to issue purchase orders.

A method of recording purchase orders is important. A numerical system often is used, and copies of the purchase orders are kept in numerical order in the purchaser's files.

RECEIVING FOOD

The person responsible for receiving food and supplies must be familiar with the specifications. Proper receiving practices ensure that the product received matches the product specified. It is the responsibility of the purchasing party to inspect products for condition and to check them against the invoice to be sure quantity and quality ordered were delivered. The unit price quoted should be compared with the unit price delivered.

Any food specified by weight should be weighed as it is delivered. Fresh and frozen products should be inspected for quality, the temperature of meat and dairy products assessed, and each item counted. Only after making sure all goods received are in satisfactory condition should the delivery slip from the purveyor be signed. *Deliveries never should be accepted without verifying weight or count, quality, and price.*

Purveyors sometimes have to make substitutions. But before accepting any substitute, the purchaser must be sure it will meet the needs of the menu and its quality, quantity, and weight should be checked. Items that are damaged or seem to have been at a higher temperature earlier in handling should be rejected and reported to the purveyor immediately. These items should be set aside to be returned or not accepted for delivery. Products should be refused if they

- Are not ordered
- Are not of required quality
- Are not the price that was quoted
- Are not delivered on a timely basis

In many operations, production people are assigned the task of receiving the orders, so delivery times should be specified. It is important to receive deliveries when there is time to check them and get them into appropriate storage immediately. Delivery during mealtime may mean the products will be checked quickly and inadequately.

STORING FOOD

Proper storage of food from the time it is received from the purveyor until it is served is important in cost and quality control. After items have been checked and accepted, they should be unpacked and immediately placed in the proper storage areas in order to preserve and safeguard quality and quantity.

There is more to storage than just having a place to put things. Storage areas must have the right light, temperature, ventilation, air movement, and security. Vendors should never be allowed to deliver directly to the storage area without adequate supervision. Ideally, storage areas should be located between the receiving and production areas.

Every food has a *shelf life,* which is the amount of time that the product reasonably can be expected to maintain its quality if stored properly. It includes the time the food is processed to the time it can no longer be used for eating.

Storage decisions are based on whether the food is perishable or nonperishable. Perishable foods, such as fresh fish and ground meats, are easily invaded by bacteria, mold, and yeast and spoil quickly. Nonperishable foods, such as flour, sugar, and canned goods, can be kept for much longer periods of time if stored correctly. These foods have low water content and must be kept dry.

The principles of storing food include the following:

- Similar items are grouped together. For example, all cans should be stored in one place and grouped by the type of food they contain. The products usually then are arranged alphabetically, such as beans, beets, corn, etc.
- Foods that absorb odors are stored away from those that give off odors.
- Stock is rotated. As new items come in, the older items should be moved to the front to be used first. Checking the refrigerator contents daily will help ensure that leftover food or items with broken packages are used quickly.
- Storage areas are kept clean and sanitized. A frequent cleaning schedule should be in place and all spills wiped up immediately.
- Safety is an important consideration. Only sturdy stepladders should be used to reach high shelves. Heavier items should be stored near the floor.
- All foods and supplies are stored off the floor.
- Shelves allow for air circulation.
- Storage area is secure. It should not be left open when unattended. Proper security must be provided between storage and receiving. A different employee from the one receiving should store and control the inventory, if possible.

The Dry-Storage Area

Dry storage is used for nonperishables that do not require refrigeration. The dry-storage area must be dry, cool (not over 70 F), properly ventilated, and controlled for humidity. Darkness and dampness allow molds to grow. Many staples, such as flour, sugar, and rice, will lose their quality in a damp area. Foods that are kept in dry storage often attract insects and rodents and thus must be stored in tightly covered containers.

Foods such as potatoes, onions, and squash may be kept at 50–60 F. Sometimes they are placed in a dry storage area if the temperature there remains cool. Crisp foods, such as crackers, need warmer temperatures of 65 F.

Paper products may be stored in this area, but cleaning supplies *never* should be stored with food because they might be confused with food items. Storage areas should be kept free from clutter. Empty boxes should be removed from the area. State and federal governments have standards for storage in health care facilities.

Refrigerated or Freezer Storage

Perishable food items must be stored properly to keep their quality and remain bacteria-free and safe for eating. Fresh and frozen products should be placed in the refrigerator or freezer immediately after delivery and kept at these temperatures until used. Here are recommended temperatures:

Perishable Food	Storage Temperature
Fresh fruits and vegetables	40–45 F
Meat, poultry, dairy products, eggs	32–40 F
Frozen products	0 to −20 F

When possible, separate refrigerators should be used for fruits and vegetables and for meat, poultry, fish, dairy products, and eggs. This procedure allows for an optimum temperature for each and minimizes the cross-contamination of fresh fruits and vegetables by poultry and meat products.

The circulation of cold air dries out (dehydrates) foods in the refrigerator or freezer. The lower the temperature, the faster foods will dehydrate. Therefore, all foods must be kept covered or wrapped. Because of their high water content, fruits and vegetables may freeze if the refrigerator temperature is too low. Fruits and vegetables must be checked often to remove any decaying portions to prevent further spoilage.

All refrigerators and freezers should have thermometers. A remote reading thermometer placed outside the low-temperature storage unit, which permits reading the temperature without opening the door is the best choice. Temperatures must be checked on a regular schedule, at least twice daily, and any problems handled as soon as they are discovered.

Inventory Control

An *inventory* is a listing of the food and supplies on hand. Accurate records are necessary for inventory control and to provide a basis for purchasing additional items as well as for cost control and to prevent theft. Each foodservice organization needs a system of keeping records up to date.

Perpetual or physical inventory systems are generally used in foodservice organizations. The *perpetual inventory* is the process of maintaining a continuous record of all purchases and issues. This inventory system is used most often in large foodservice organizations that carry bigger inventories. Perpetual inventory usually is restricted to products in the dry storage and frozen storage areas.

Physical inventory involves a periodic counting of products in all storage areas. Generally, inventories are counted once each month, usually on the last working day of the month. The process should involve two people, one counting the number of units and one recording the information on a recording form.

If a perpetual inventory system is used, the physical system should also be done on a monthly basis to verify the accuracy of the perpetual system. Any discrepancies should be identified to find out the reason why they have occurred.

SUMMARY

Food purchasing, receiving, and storing are important functions of any foodservice operation in order to maintain quality food products. The objective of the food purchaser is to ensure that the right product is available in the amount needed at the time needed. The food purchaser must be familiar with the government regulations that have been established to make sure quality food is available for purchase. Quality grades have been established for many products.

Many purchasing decisions must be made regarding methods of purchasing and the quality and quantity needed. Written specifications should be used to get the quality of product needed by the foodservice operation. Specifications are used both for developing purchase orders and for issuing bid requests.

Proper receiving practices should be used to ensure that the quality specified is the quality received. Weight or quantity, quality, and price should be verified. Appropriate storage areas need to be identified for food products and supplies to maintain quality. Proper security between receiving and storage must be provided.

The two inventory systems used in foodservice organizations are perpetual and physical. They are usually used together.

LEARNING ACTIVITIES

Activity 1: Reviewing the Purchasing System

A. List the important components of an effective purchasing system.

_____

_____

_____

B. How does competitive buying differ from one-stop buying?

_____

_____

_____

C. What are the disadvantages of choosing a purveyor only on the basis of the lowest cost?

_____

_____

D. What is the significance of a purchaser accepting an ice cream cake as a gift from a purveyor?

_____

_____

_____

E. What are disadvantages of buying only on price?

_____

_____

_____

Activity 2: Evaluating an Actual Foodservice Purchasing and Storage System

1. Use the checklist to evaluate the purchasing, receiving, and storage procedures of your foodservice organization.

Activity 2, Step 1

Procedure	Yes	Sometimes	No
A. Deliveries from purveyors are checked for quantity or weight, quality, and price.			
B. Unacceptable items are returned to the purveyor.			
C. Similar items are groups in storage.			
D. Stock is rotated using the oldest first.			
E. All food is stored off the floor.			
F. A cleaning schedule is in place.			
G. Storage areas are cleaned on a regular basis.			
H. Storage areas have adequate lighting.			
I. Storage areas have adequate ventilation.			
J. Dry storage areas are between 50 and 70 F all year around.			
K. All foods in refrigerator and freezer are tightly covered or wrapped.			
L. Each refrigerator and freezer has a thermometer.			
M. Storage areas are clear of clutter.			
N. Cleaning supplies are not located next to food items.			
O. Sturdy stepladders are available for reaching high areas.			
P. Storage areas are locked when unattended.			
Q. A system for keeping up-to-date records is in place.			

2. Review the guidelines for receiving and storing products in your facility.

   A. If no guidelines are in place, list some that would be useful.

   _____

   _____

   _____

   _____

   B. Are there any opportunities for dishonesty? Why or why not?

   _____

   _____

   _____

FOOD PURCHASING, RECEIVING, AND STORING

C. Complete the chart on storing foods (pages 211 and 212). List all the conditions necessary to keep the food at its highest quality.

Activity 2, Step 2C

Food	Storage Area	Temperature	Other Conditions
Flour			
Eggs			
Dairy products			
Fresh meat and poultry			
Rice			
Canned goods			
Apples			
Potatoes			
Frozen vegetables			

Food	Storage Area	Temperature	Other Conditions
Powdered milk			
Bananas			
Stainless steel cleaner			
Lettuce			

D. Write specifications for your facility for the following food items.

Canned peaches that will be used for dessert. (Two halves will be served for the client's dessert.)

_____

_____

Frozen cut green beans. (Each client will receive ½ cup.)

_____

_____

REVIEW QUESTIONS

True or False

1. Only the person purchasing the food needs to be aware of the written specifications.

    A. True
    B. False

2. It usually is easier to reject items before they are accepted than after the delivery slip has been signed.

   A. True
   B. False

3. The quantity or weight of all items must always be checked as a receiving function.

   A. True
   B. False

4. Both the purchaser and the purveyor must know the federal, state, and local regulations.

   A. True
   B. False

Multiple Choice

5. This is a statement of what is contained in a food product.

   A. Standard of quality
   B. USDA quality grade
   C. Fair Labeling Act
   D. Standard of identity

6. Which of the following is most important when making purchasing decisions?

   A. The nutritional needs of the clients
   B. The skills of the personnel
   C. The food budget
   D. The amount of storage space
   E. All of the above

Food Preparation Terms

Bake: Cook by dry heat in an oven or on heated metals. The term **roast** is used when it is applied to meat in uncovered containers.
Baste: Moisten food while cooking to add flavor and prevent drying of the surface. The liquid may be melted fat, meat drippings, water, sauce, or juice.
Beat: Mix ingredients thoroughly, usually in a bowl, using an over-and-under motion.
Blanch: Preheat in boiling water or steam. This method is used to inactivate enzymes and shrink some foods for canning, freezing, or drying. It is also used to aid in removal of skins from nuts, fruits, and some vegetables.
Blend: Thoroughly mix two or more ingredients.
Braise: Cook slowly in a covered utensil in a small amount of liquid at a low temperature. (Meat may or may not be browned in a small amount of fat before braising.)
Bread: Dip in seasoned flour, or fine bread, cracker, or cereal crumbs, and then into moisture, such as egg, milk, or egg plus liquid, and roll in crumbs.
Broil: Cook by direct heat.
Brown: Bake, broil, fry, or toast a food until the surface is browned.
Brush: Spread a coating, such as melted butter, on top of food with a brush or paper towel.
Caramelize: Heat sugar or foods containing a high percentage of sugar until a brown color and caramel flavor develop.
Chill: Allow a food to become thoroughly cold in a refrigerator.
Chop: Cut food into small pieces.
Coat: Cover food evenly with flour, sugar, crumbs, or nuts. Sometimes the food is dipped in an egg-milk mixture before coating.
Cool: Let stand at room temperature until the food is no longer warm to the touch.
Cream: Mix together one or more foods until soft, smooth, and creamy.
Cube: Cut into cubes.
Cut in: Chop solid fat into dry ingredients with a pastry blender or knives until the fat is finely divided.
Dice: Cut into cubes.
Dilute: Add water to another liquid.
Dissolve: Mix a solid ingredient with a liquid until they form a solution.
Dot: Partially cover with small particles, such as butter.
Dredge: Coat with flour or other fine substance.
Flake: Break up into small pieces with a fork.
Flour: Roll in seasoned flour; coat with flour and shake off excess.
Fold: Combine ingredients with two motions, cutting vertically through the mixture and then turning it over by sliding the mixing implement across the bottom of the bowl.
Fricassee: Brown small pieces of meat, usually fowl or veal, in a small amount of fat and then stew or braise.
Fry: Cook in hot fat.

Glaze: Coat with a thin sugar syrup that has been cooked to the crack stage.
Grate: Rub food on a grater to make small pieces.
Grill: Cook by direct heat, using a griddle. Also refers to use of a gas or charcoal grill.
Julienne: Cut food into narrow lengthwise strips.
Knead: Manipulate with a pressing motion accompanied by folding and stretching.
Lard: Insert small strips of fat into, or place on top of, uncooked lean meat or fish to give flavor and prevent dryness.
Marinate: Let food stand in a liquid mixture called a marinade, which may be an oil-acid mixture, vinegar, lemon juice, sour cream, etc.
Melt: Change a solid food to liquid by heating.
Mince: Cut into very fine pieces, much smaller than those achieved by chopping or dicing.
Mix: Combine two or more ingredients by beating or stirring.
Panbroil: Cook uncovered on a hot surface or a shallow pan and pour off fat as it accumulates.
Panfry: Cook in a small amount of fat in a shallow pan.
Parboil: Partially cook a food in boiling water prior to further cooking or other method of preparation.
Pare: Cut a very thin layer of peel from fruits or vegetables.
Peel: Pull off the rind or outer covering of certain fruits or vegetables.
Poach: Cook food in a hot liquid, using care to retain its form.
Puree: Strain or blend to a paste or semi-liquid.
Reconstitute: Restore concentrated foods to their original state by adding a liquid. Also called **rehydrate**.
Roast: Cook uncovered by dry heat in an oven or in heated metals. The term is usually applied to meat.
Sauté: Cook in a small amount of fat.
Scald: Heat a liquid to a point just below boiling.
Scallop: Bake food, usually covered with a liquid or sauce and crumbs. Also called **escallop**.
Score: Make shallow lengthwise and crosswise slits on the surface.
Sear: Brown the surface of meat, using intense heat for a short time.
Shred: Tear food into long, thin pieces; grate food coarsely on a grater.
Sift: Put dry ingredients through a fine sieve.
Simmer: Cook in a liquid in which bubbles form slowly and break below the surface, just below the boiling point.
Skim: Remove the top layer from a liquid, such as fat from gravy.
Steam: Cook in steam, with or without pressure.
Steep: Soak substance in liquid just below the boiling point for softening, coloring, or extracting flavor.
Stew: Simmer in small amount of liquid.
Stir: Mix food materials with a circular motion to blend mixture to a uniform consistency.
Toast: Brown in an oven broiler or toaster.
Toss: Tumble ingredients, such as for a salad, very lightly with a spoon and fork.
Weeping: Leakage or collection of fluid between the filling and meringue due to failure to denature the protein of the egg white.
Whip: Beat rapidly to increase volume by incorporating air.

Answers to Chapter Review Questions

1. Recipe Standardization
   1. A
   2. B
   3. B
   4. B
   5. D
   6. C
2. Fruits
   1. B
   2. A
   3. B
   4. B
   5. A
   6. A
   7. B
   8. C
   9. B
   10. 1C, 2B, 3E, 4F, 5G, 6H
3. Vegetables
   1. B
   2. A
   3. A
   4. A
   5. B
   6. B
   7. A
   8. A
   9. C
   10. B
   11. B
   12. B
   13. 1C, 2F, 3D, 4I, 5B, 6G, 7E, 8H
4. Salads
   1. A
   2. B
   3. B
   4. A
   5. A
   6. A
   7. C
   8. C
   9. B
   10. C
   11. B
   12. C
   13. 1C, 2D, 3E, 4F, 5A, 6G, 7H
5. Starches, Sauces, Soups, Cereals, and Pastas
   1. A
   2. A
   3. B
   4. A
   5. B
   6. B
   7. B
   8. B
   9. D
   10. A
   11. B
   12. B
6. Milk and Cheese
   1. C
   2. B
   3. C
   4. C
   5. D
7. Meats, Poultry, Fish, and Entrees
   1. B
   2. B
   3. D
   4. D
   5. D
   6. A
   7. D
   8. B

ANSWERS TO CHAPTER REVIEW QUESTIONS

8. Eggs and Egg Products
   1. A
   2. B
   3. B
   4. C
   5. B
   6. E
   7. B
   8. C
   9. D

9. Doughs, Batters, and Pastries
   1. B
   2. D
   3. C
   4. C
   5. C
   6. B
   7. C
   8. C

10. Beverages and Convenience Foods
    1. B
    2. A
    3. D
    4. B
    5. C
    6. B
    7. D
    8. B

11. Microwave Cooking
    1. Suggested answers:
       A. Arcing: A static spark caused by two pieces of metal being close together or too close to the oven wall during microwave cooking. Arcing is commonly caused by foil, metal rims on china, and metal twist ties.
       B. Wattage: A measurement of the amount of cooking power created by the microwave oven. The number of watts generated indicates the wattage. The greater the wattage, the faster foods will cook.
       C. Shielding: Covering part of the food to make it cook more slowly or to prevent it from cooking, such as covering the corners of a square pan with foil.
       D. Conduction: The heating of an item by coming in contact with something hot. The interior of large items cooked by microwaving are cooked by conduction, heated by being in contact with the microwave-heated outside surfaces. The heating of food in a conventional oven, convection oven, and on a range top is also by conduction.
       E. Density: A way to describe the solidness of foods. Dense foods have few air spaces while less dense foods have many air spaces and are light and airy.
    2. Suggested answers:
       A. Because cooking time will vary, depending on the temperature of the food before cooking.
       B. Microwaved foods cook in the center last. The outside can appear thoroughly cooked yet the inside may still be undercooked.
       C. Foods heat unevenly because areas with high moisture, fat, or sugar heat faster than other areas. This may cause some parts of food to be too hot to serve while other parts are still too cool.

12. Menu Planning
    1. A
    2. B
    3. B
    4. B
    5. D
    6. A
    7. E

13. Food Purchasing, Receiving, and Storing
    1. B
    2. A
    3. A
    4. A
    5. D
    6. E

References

American Egg Board. 1991. *The Incredible Edible Egg*. Park Ridge, Ill.: American Egg Board.
Charley, H. 1986. *Food Science*. New York: Macmillan.
Cummings, L., and L. Kotschevar. 1989. *Nutrition Management for Foodservices*. New York: Delmar.
Gisslen, W. 1989. *Professional Cooking*. New York: Wiley.
_____. 1990. *Professional Baking*. New York: Wiley.
Hullah, E. 1984. *Cardinal's Handbook of Recipe Development*. Don Mills, Ont., Can.: Cardinal Kitchens.
Klein, B., E. Matthews, and C. Setser. 1984. *Foodservice Systems: Time and Temperature Effects on Food Quality*. North Central Regional Research Publication No. 293, Illinois Bulletin 779. Urbana, Ill.: University of Illinois Agricultural Experiment Station.
Kotschevar, L. 1988. *Standards, Principles and Techniques in Quantity Food Production*. 4th ed. New York: Van Nostrand Reinhold.
Kowtaluk, H., and A. Kopan. 1986. *Food for Today*. 3d ed. Manchester, Mo.: Glencoe.
Largen, V. 1988. *Guide to Good Food*. South Holland, Ill.: The Goodheart.
Mario, T. 1978. *Quantity Cooking*. Westport, Conn.: AVI Publishing.
Matthews, R., and Y. Garrison. 1975. *Food Yields Summarized by Different Stages of Preparation*. Agriculture Handbook No. 102. Washington, D.C.: U.S. Department of Agriculture.
McWilliams, M. 1985. *Food Fundamentals*. 4th ed. New York: Wiley.
_____. 1989. *Foods: Experimental Perspectives*. New York: Macmillan.
Mizer, D., M. Porter, and B. Sonnier. 1987. *Food Preparation for the Professional*. New York: Wiley.
National Association of Meat Purveyors. 1988. *The Meat Buyer's Guide*. Chicago: National Association of Meat Purveyors.
Pauli, E. 1989. *Classical Cooking the Modern Way*. New York: Van Nostrand Reinhold.
Penfield, M., and A. Campbell. 1990. *Experimental Food Science*. 3rd ed. New York: Academic Press.
Potter, N. 1968. *Food Science*. Westport, Conn.: AVI Publishing.
Powers, J. 1979. *Basics of Quantity Food Production*. New York: Wiley.
Ray, M., and E. Lewis. 1986. *Exploring Professional Cooking*. 3rd ed. Peoria, Ill.: Bennett Publishing.
Schmidt, A. 1989. *Chef's Book of Formulas, Yields, and Sizes*. New York: Van Nostrand Reinhold.
Shugart, G., and M. Molt. 1989. *Food for Fifty*. New York: Macmillan.
Spears, M., and A. Vaden. 1985. *Foodservice Organizations*. New York: Wiley.
West, B., and L. Wood. 1988. *Foodservice in Institutions*. 6th ed. New York: Macmillan.

Index to Recipes

Apples, Baked, 34
Beef Roast, Round, 118
Biscuits, Shortcake, 89
Cake, Devil's Food, 159
Chicken, Oven-fried, 6
Chicken Newburg, 122
Chowder, Vegetable, 100
Cobbler, Cherry
  for 25 portions, 88
  for 100 portions, 77
Cookies, Chocolate Chip, 160
Eggs, Hard-cooked, 139
Meat Loaf, 120
Milk, Nonfat Dry, Reconstituting, 96
Muffins, Blueberry, 155

Pie, Cream, 76
Rice, Oven-cooked, 83
Salad, Apple Wedge, 64
Salad, Tossed, 68
Salad Dressing, Italian, 67
Salad Liners, 60
Sauce, Mushroom, 98
Sauce, Tomato, 78
Sauce, Veloute, 97
Sauce, White
  using fresh milk, 96
  using nonfat dry milk, 97
Soup, Cream of Mushroom, 99
Vegetable Medley, Fresh, 53